Natural Environment Research Council

Institute of Terrestrial Ecology

A Field Key for Classifying British Woodland Vegetation Part 1

R. G. H. Bunce
Institute of Terrestrial Ecology
Merlewood Research Station
Grange-over-Sands
Cumbria LA11 6JU

Printed in Wales by
Cambrian News (Aberystwyth) Ltd.

Published in 1982 by
Institute of Terrestrial Ecology
68 Hills Road
Cambridge
CB2 1LA

ISBN 0 904282 68 6

The Institute of Terrestrial Ecology (ITE) was established in 1973, from the former Nature Conservancy's research stations and staff, joined later by the Institute of Tree Biology and the Culture Centre of Algae and Protozoa. ITE contributes to, and draws upon, the collective knowledge of the fourteen sister institutes which make up the **Natural Environment Research Council,** spanning all the environmental sciences.

The Institute studies the factors determining the structure, composition and processes of land and freshwater systems, and of individual plant and animal species. It is developing a sounder scientific basis for predicting and modelling environmental trends arising from natural or man-made change. The results of this research are available to those responsible for the protection, management and wise use of our natural resources.

One quarter of ITE's work is research commissioned by customers, such as the Department of the Environment, the European Economic Community, the Nature Conservancy Council and the Overseas Development Administration. The remainder is fundamental research supported by NERC.

ITE's expertise is widely used by international organisations in overseas projects and programmes of research.

CONTENTS

INTRODUCTION

    Procedure in the field

    Worked example

    Type descriptions

    1. Vegetation

    2. Environment

    3. General descriptions

    Acknowledgements

    References

THE KEY

DETAILED DESCRIPTIONS OF TYPES OF WOODLAND VEGETATION
including distribution maps and colour photographs.

INTRODUCTION

Although only about 9% of Britain is afforested, the country has a wide variety of native woodlands, including broadleaved species (eg oak, birch and ash) and coniferous species (Scots pine and yew); 95% of such woodlands are predominantly broadleaved, whereas under 5% are predominantly coniferous.

These woodlands have been affected directly and indirectly by the activities of man, with changes in the balance of tree species having secondary effects on (i) amounts of light reaching the forest (woodland) floor, (ii) soil properties, etc. With these changes and naturally occurring small-scale variations attributable to bedrock geology and topography, it is not surprising that complex arrays of ground vegetation have developed. For example, assemblages of herbaceous species in stands of oak and birch can differ greatly. It is possible to identify woodlands by their arrays of trees—a fairly coarse separation. However, it is also possible to subdivide oak and Scots pine woodlands by their associated assemblages of ground vegetation, the latter being much more meaningful ecologically.

Because of this complexity in woodland structure attributable to trees and ground vegetation, a survey was initiated in 1969 to provide data for the production of an integrated system of **classification.**

A survey was made of 103 cartographically-defined woodlands, which were selected to encompass the range of variation within Britain. The selection was based on species lists prepared by the staff of the former Nature Conservancy who had inspected 2,463 woodland sites (10% of the British complement).

At each of the 103 woodlands, observations were made of the vegetation within 16 randomly distributed plots, each 200 m^2 (Bunce & Shaw 1973). This relatively large plot size was chosen on the basis of the experience of continental phytosociologists who found that it was necessary to minimize the extreme variation associated with often major variations in light penetration over relatively short distances. This booklet, Part I, is concerned with the **classification of vegetation within the separate plots,** and includes colour photographs of each of the 32 plot types. A second booklet, Part II, will consider the integration of plot data into woodland descriptions.

For each plot, a list was made of (i) vascular plants, and (ii) bryophytes growing on the ground. Additionally, diameters of all trees, at breast height, were recorded, together with a standard list of habitat attributes (Bunce & Shaw 1973).

The types of woodland assemblages identified by the Key described in this booklet were generated by a numerical technique, indicator species analysis (Hill *et al.* 1975). Unlike most systems for classifying vegetation, the use of indicator species analysis does not presuppose the existence of dominants—in the first instance, all the species are treated equally. Furthermore, use of the present Key does not require previous experience of vegetation classification, and, because the random samples were situated throughout woodlands, whether in glades or not, the approach applies to all parts of a woodland.

The present classification differs from other woodland classifications in a number of respects:

1. It is based on a survey using a standardised sampling system, with randomly placed plots, covering a wide range of British woodlands.
2. The classification is minimally dependent on subjective judgements.
3. The classification depends, at one and the same time, on the arrays of (i) trees and (ii) other plant species (understorey species and ground vegetation).

The nomenclature for Latin and English names of vascular plants follows Clapham *et al.* (1962) and for bryophytes, Watson (1955). These authorities were thought to be those most likely to be used by potential readers. Specimens of all species, identified by the survey teams, were collected, and subsequent decisions on combining species were made in order to ensure a uniform and consistent standard of identification. Species of oak, willow and birch are not separated; the species of *Taraxacum* and *Leontodon* are likewise grouped. *Rubus fruticosus, Dryopteris dilatata* and *Dryopteris filix-mas* are treated as aggregates. *Viola reichenbachiana* is included under *V. riviniana, Cardamine hirsuta* under *C. flexuosa,* and *Poa nemoralis* under *P. trivialis. Epilobium montanum* may also include hybrids and possibly some related species. The various relatives of *Ulmus procera* and *U. carpinifolia* are grouped together.

Procedure in the field
The Key provides a means of sorting the different assemblages, using a relatively restricted number of indicator species. The accuracy of the method in assigning new individuals to their appropriate types is considered to be high in relation to either observer or sampling errors in the 'field'. However, if exceptional plots are being observed, eg amongst scrub on sea cliffs, and/or if plots are being examined in exceptional circumstances, eg a drought year, problems could arise.

The survey method described by Bunce and Shaw (1973) should be modified as little as possible, but data collected from comparable areas may also be used, although with less confidence. Having located 200 m^2 plots, a list of (i) vascular plants (and their percentage ground cover in 5% categories) and (ii) ground bryophytes should be prepared, and measurements taken of tree diameters at breast height which, in the absence of other records, give some indication of age (see Figure 1). These tasks are simplified by the progressive search of nested quadrats 2 × 2 m, 5 × 5 m, 7 × 7 m, 10 × 10 m, and finally, 14 × 14 m, which are conveniently located by markers attached to the diagonals of a 14 × 14 m square (see Figure 2). To add interest and to facilitate ecological understanding, the habitat attributes listed in Figure 3 should be recorded.

Figure 1. Data sheet for recording diameters of trees, saplings and shrubs, using the method described by Bunce and Shaw (1973)

TREE, SAPLING AND SHRUB DATA

Site No. 11   Plot No. 8   Recorder SKM/JMS   Date 14/7/71   Ht (m)

Q No.	Species	Diameter Breast Height (cm)						
1								
T R E E S	Field maple	6	3	9				
	Oak	36	7					
S A P S	English elm	4						
S H R B	Spindle	1						
	Elder	3						
2								
T R E E S	Oak	25	4ᴰ	3ᴰ	36	36		
	Field maple	7						
3								
T R E E S	English elm	8	10	4	3	3		
	Gean	11						
S A P S	Field maple	2						
	Gean	4						
S H R B	Hawthorn	2	2					
	Spindle	3	2					
4								
T R E E S	Gean	7	6					
	Oak	30	48					

D, dead

Figure 2. Layout of ground flora quadrats.
Distance string position from centre = ½ diagonal
2 m quadrat (4 m²) = 1·42 m
1 Q*= 25 m² (5·00 × 5·00 m) = 3·54 m
2 Q = 50 m² (7·07 × 7·07 m) = 5·00 m
3 Q = 100 m² (10·0 × 10·0 m) = 7·07 m
4 Q = 200 m² (14·14 × 14·14 m) = 10·0 m

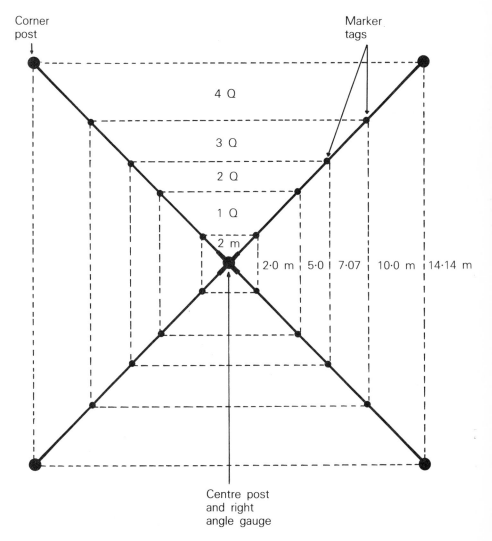

* 1Q, 2Q, 3Q and 4Q, designation of quadrats illustrated in diagram

7

Figure 3. Data sheet for recording plot description and habitats, as described by Bunce and Shaw (1973)

PLOT DESCRIPTION AND HABITATS

Site No. 11   2 Plot No. 8   3 Recorder SKM/JMS   4 Date 14/7/71
5 Slope 34° or %   6 Aspect °Mag.

A TREES—MANAGEMENT
7 Cop. stool   8 Singled cop.   9 Rec. cut. cop.   10 Stump hard. new
~~11 Stump hard. old~~   12 Stump con. new   13 Stump con. old   14

B TREES—REGENERATION

15 Alder	16 Ash	17 Aspen	18 Beech
19 Birch	~~20 Hawthorn~~	21 Hazel	22 Holly
23 Hornbeam	24 Lime	25 Oak	26 Rowan
27 Rhododendron	28 Sweet chestnut	29 Sycamore	30 Wych elm
31 Other hrwd.	32 Scots pine	33 Yew	34 Other con.

C TREES—DEAD (= HABITATS)

35 Fallen brkn.	~~36 Fallen uprtd.~~	37 Log. v. rotten	38 Fall. bnh. >10cm
39 Hollow tree	40 Rot hole	41 Stump <10cm	~~42 Stump >10cm~~

D TREES—EPIPHYTES AND LIANES

43 Bryo. base	44 Bryo. trunk	45. Bryo. branch	46 Lichen trunk
47 Lichen branch	48 Fern	~~49 Ivy~~	50 Macrofungi

E HABITATS—ROCK

~~51 Stone <5cm~~	~~52 Rocks 5-50cm~~	53 Boulders >50cm	54 Scree
55 Rock outcp. <5m	56 Cliff >5cm	57 Rock ledges	58 Bryo. covd. rock
59 Gully	60 Rock piles	61 Exp. grav/sand	~~62 Exp. min. soil~~

F HABITATS—AQUATIC

63 Sml. pool <1m²	64 Pond 1-20m²	65 Pond/lake>20m²	66 Strm/riv. slow
67 Strm/riv. fast	68 Aquatic veg.	69 Spring	70 Marsh/bog
71 Dtch/drain dry	72 Dtch/drain wet	73	74

G HABITATS—OPEN

75 Gld. 5-12m	76 Gld.>12m	77 Rky. knoll <12m	78 Rky. knoll >12m
~~79 Path <5m~~	80 Ride >5m	81 Track non-prep	82 Track metalled

---

H HABITATS—HUMAN

83 Wall dry	84 Wall mortared	85 Wall ruined	86 Embankment
87 Soil excav.	88 Quarry/mine	~~89 Rubbish dom.~~	90 Rubbish other

---

I HABITATS—VEGETATION

91 Blkthorn. thkt.	92 Hawthorn thkt.	93 Rhodo. thkt.	94 Bramble clump
95 Nettle clump	96 Rose clump	97 W.herb clump	98 Umbel. clump
99 Bracken dense	100 Moss bank	101 Fern bank	102 Grass bank
103 Leaf drift	104 herb veg. >1m	105 Macfungi. soil	~~106 Macfungi. wood~~

J ANIMALS (mainly signs)

107 Sheep	108 Cattle	109 Horse/pony	110 Pigs
111 Red deer	112 Other deer	~~113 Rabbit~~	114 Badger
115 Fox	116 Mole	117 Squirrel	118 Anthill
119 Corpse/bones	120 Spent ctrdgs.	121	122

COMMENTS

Worked example

The following species were listed from a plot, 200 m², in a wood with birch above Lochan an Draing in Wester Ross. No distinction should be made when their contributions as indicators are considered—presence alone is sufficient. Species with no English name attached are bryophytes and follow the flowering plants at all stages in the text.

Quadrat 1
The smallest
(4 m²) of the
nested quadrats.

Deschampsia flexuosa (wavy hair-grass)
Anthoxanthum odoratum (sweet vernal-grass)
Oxalis acetosella (wood-sorrel)
Pteridium aquilinum (bracken)
Agrostis tenuis (common bent-grass)
Holcus mollis (creeping soft-grass)
Galium saxatile (heath bedstraw)
Sorbus aucuparia (rowan)
Hylocomium splendens
Rhytidiadelphus squarrosus
Polytrichum spp
Hypnum cupressiforme

Quadrat 2
25 m²

Species additional to those in Quadrat 1:
Blechnum spicant (hard-fern)
Corylus avellana (hazel)
Betula spp (birch)
Vaccinium myrtillus (bilberry)
Pleurozium schreberi
Pseudoscleropodium purum
Dicranum scoparium
Thuidium tamariscinum

Quadrat 3
50 m²

Species additional to those in Quadrats 1 and 2:
Potentilla erecta (common tormentil)
Erica cinerea (bell-heather)
Isothecium myosuroides
Dicranum majus

Quadrat 4
100 m²

Species additional to those in Quadrats 1, 2 and 3:
Dryopteris dilatata (broad buckler-fern)
Calluna vulgaris (ling)
Carex pilulifera (pill-headed sedge)
Plagiothecium undulatum
Tetraphis pellucida
Lepidozia reptans

Quadrat 5
200 m²

Species additional to those in Quadrats 1, 2, 3 and 4:
Hypericum pulchrum (slender St. John's wort)
Luzula multiflora (many-headed woodrush)
Agrostis canina (brown bent-grass)

Having prepared the species list, it is now appropriate to consider STEP ONE of the Key (Page 21). For convenience, species in the left-hand column are printed in capital letters, while those in the right-hand column are printed in lower case. If a species from the left-hand column is present, it scores −1, ie it counts negatively; if a species from the right-hand column is present, it scores +1, ie it counts positively. Before keying out the above species list, it is advisable to consider some hypothetical examples to indicate how the balance between negative and positive species is achieved.

STEP ONE

Negative	Positive
CIRCAEA LUTETIANA (ENCHANTER'S NIGHTSHADE)	Anthoxanthum odoratum (sweet vernal-grass)
FRAXINUS EXCELSIOR (ASH)	Deschampsia flexuosa (wavy hair-grass)
GEUM URBANUM (HERB BENNET)	Galium saxatile (heath bedstraw)
MERCURIALIS PERENNIS (DOG'S MERCURY)	Pteridium aquilinum (bracken)
EURHYNCHIUM PRAELONGUM	Polytrichum spp

Score −1 or less (ie −1, −2, −3, . . .), go to STEP TWO
Score 0 or more (ie 0, +1, +2, . . .), go to STEP SEVENTEEN

Five hypothetical combinations of indicator species are shown below, with their appropriate scores.

a. CIRCAEA LUTETIANA (ENCHANTER'S NIGHT-SHADE) −1
FRAXINUS EXCELSIOR (ASH) −1
GEUM URBANUM (HERB BENNET) −1

= −3 (i.e. Go to STEP TWO)
(less than the threshold of −1)

b. CIRCAEA LUTETIANA (ENCHANTER'S NIGHT-SHADE) −1
FRAXINUS EXCELSIOR (ASH) −1
GEUM URBANUM (HERB BENNET) −1
Galium saxatile (heath bedstraw) +1
Pteridium aquilinum (bracken) +1

= −3+2 = −1 (i.e. Go to STEP TWO)
(equal to the threshold of −1)

c. FRAXINUS EXCELSIOR (ASH) −1
Pteridium aquilinum (bracken) +1

= −1+1 = 0 (i.e. Go to STEP SEVENTEEN)

10

d. *FRAXINUS EXCELSIOR*   −1 ⎫
       (ASH)   ⎪
    Pteridium aquilinum   +1 ⎬ = −1+2 = +1   (i.e. Go to STEP
       (bracken)   ⎪   SEVENTEEN)
    Galium saxatile   +1 ⎭
       (heath bedstraw)

e. *Anthoxanthum odoratum*   +1 ⎫
       (sweet vernal-grass)   ⎪
    Deschampsia flexuosa   +1 ⎬ = +3   (i.e. Go to STEP
       (wavy hair-grass)   ⎪   SEVENTEEN)
    Galium saxatile   +1 ⎭
       (heath bedstraw)

Now consider the data for the wood with birch at Lochan an Draing. On inspecting
the species list for the indicator species in STEP ONE, the following should be
identified:

STEP ONE
(Page 21)

Negative	*Positive*	
	Anthoxanthum odoratum	+1
	(sweet vernal-grass)	
	Deschampsia flexuosa	+1
	(wavy hair-grass)	
	Galium saxatile	+1
	(heath bedstraw)	
	Pteridium aquilinum	+1
	(bracken)	
	Polytrichum spp	+1

Total score +5, therefore proceed to STEP SEVENTEEN

STEP SEVENTEEN
(Page 29)

By referring to Page 29 and the species list, the following indicators were identified
for this step:

Negative	*Positive*	
	Agrostis canina	+1
	(brown bent-grass)	
	Anthoxanthum odoratum	+1
	(sweet vernal-grass)	
	Galium saxatile	+1
	(heath bedstraw)	
	Potentilla erecta	+1
	(common tormentil)	
	Rhytidiadelphus squarrosus	+1

11

Total score +5, therefore proceed to STEP TWENTY-FIVE

STEP TWENTY-FIVE
(Page 33)

By referring to Page 33 and the species list, the following indicators were identified for this step:

Negative		Positive	
CALLUNA VULGARIS (LING)	−1		
DESCHAMPSIA FLEXUOSA (WAVY HAIR-GRASS)	−1		
SORBUS AUCUPARIA (ROWAN)	−1		
VACCINIUM MYRTILLUS (BILBERRY)	−1		

Total score −4, therefore proceed to STEP TWENTY-SIX.

STEP TWENTY-SIX
(Page 33)

Negative		Positive	
DRYOPTERIS DILATATA (BROAD BUCKLER-FERN)	−1	*Calluna vulgaris* (ling)	+1
OXALIS ACETOSELLA (WOOD-SORREL)	−1	*Erica cinerea* (bell-heather)	+1
		Pleurozium schreberi	+1

Total score −2+3 = +1, therefore proceed to STEP TWENTY-EIGHT

STEP TWENTY-EIGHT
(Page 35)

Negative		Positive	
AGROSTIS TENUIS (COMMON BENT-GRASS)	−1	*Agrostis canina* (brown bent-grass)	+1
GALIUM SAXATILE (HEATH BEDSTRAW)	−1	*Potentilla erecta* (common tormentil)	+1
OXALIS ACETOSELLA (WOOD-SORREL)	−1		

Total score −3+2 = −1, therefore the plot is representative of **PLOT TYPE 27, birch or oak woodland in poor, upland, freely-drained conditions.**

On inspection of the type descriptions, it will be seen that the species list of Lochan an Draing closely conforms to the array of 'constant' and 'selective' species given in the description of PLOT TYPE 27. It should also be noted that many of the species given in the complete list for this particular sample plot were not used when identifying the plot type. Because the plot types have been created by arbitrary, but reproducible divisions, some overlapping will inevitably occur. It should be noted that a species may have a negative score at one step and a positive score at another. For example, *Calluna vulgaris* (ling) was NEGATIVE at Step Twenty-five but positive at Step Twenty six.

Type descriptions
In the type descriptions, plant species are arranged in different groupings, in parallel with résumés of physical habitat attributes. These groupings need to be defined.
  Each type has been given an appellation, for example:

PLOT TYPE 1
URTICA DIOICA/RUBUS FRUTICOSUS (STINGING NETTLE/BRAMBLE)

PLOT TYPE 2
BROMUS RAMOSUS/MERCURIALIS .PERENNIS (HAIRY BROME/DOG'S MER-
CURY)

Users accustomed to traditional systems of woodland classification may find the absence of tree species in these names surprising. However, because descriptive names based solely on the relatively few kinds of trees would be unduly restricting, it was decided to depend upon the greater variety of ground vegetation species. On inspecting the description of PLOT TYPE 1, it will be seen that *URTICA DIOICA/RUBUS FRUTICOSUS* are associated with a variable canopy of ash, beech, oak and sycamore.

The names are in 2 parts. The first is the ground flora selective species (see below) that occurred in over 75% of the replicate plots of the particular type; the second is the species with highest frequency as plot dominant (see below).

1. **Vegetation**—2 divisions, ground flora and woody perennials, each of these being subdivided.

1.1 **Key species**
  1.1.1 **Constant species** These species occurred in more than 75% of plots when identifying types in the original survey of 1648 plots. These species are analogous to the 'character' species of the Zürich-Montpellier system.
  1.1.2 **Plot dominants** These species have an estimated cover of 10% or over in more than 15% of the plots when identifying types in the original survey of 1648 plots.
  1.1.3 **Selective species** To establish if a species occurred differentially, its observed frequency within a type was compared with the mean frequency over the whole series of types. The chi-square test was used to test the departure from randomness, and the 6 species with probabilities of over 99·9% are listed. These species are analogous to the 'differential' species of the Zürich-Montpellier system.
  1.1.4 **Species groups** Ground flora vascular species were classified into the following groups according to their associations one with another (see

Bunce 1977). The names for these groups summarise an interpretation of their environmental affinities; the species are in the order provided by the analysis, so that those with the closest affinities are placed next to one another. The 11 groups included:

A. (Brown earth, often gleyed soils, woodland)
   Veronica montana (wood speedwell), *Silene dioica* (red campion), *Geranium robertianum* (herb robert), *Sanicula europaea* (sanicle), *Galium odoratum* (sweet woodruff), *Bromus ramosus* (hairy brome), *Festuca gigantea* (tall brome), *Circaea lutetiana* (enchanter's nightshade), *Fragaria vesca* (wild strawberry), *Brachypodium sylvaticum* (slender false-brome), *Carex sylvatica* (wood sedge), *Urtica dioica* (stinging nettle), *Glechoma hederacea* (ground ivy), *Poa trivialis* (rough meadow-grass), *Geum urbanum* (herb bennet), *Galium aparine* (goosegrass).

B. (Brown earth, usually basiphilous soils, woodland)
   Rubus fruticosus (bramble), *Hedera helix* (ivy), *Endymion non-scriptus* (bluebell), *Mercurialis perennis* (dog's mercury), *Arum maculatum* (lords-and-ladies), *Galeobdolon luteum* (yellow archangel).

C. (Wet soils, often with organic surface, open/woodland)
   Valeriana officinalis (valerian), *Galium palustre* (marsh bedstraw), *Filipendula ulmaria* (meadow-sweet), *Angelica sylvestris* (wild angelica), *Chrysosplenium oppositifolium* (opposite-leaved golden saxifrage).

D. (Brown earth soils, often eutrophic, open/woodland)
   Veronica chamaedrys (germander speedwell), *Dactylis glomerata* (cock's-foot), *Stachys sylvatica* (hedge woundwort), *Rumex conglomeratus* (sharp dock), *Heracleum sphondylium* (hogweed), *Arrhenatherum elatius* (oat-grass), *Epilobium montanum* (broad-leaved willow-herb), *Cardamine flexuosa* (wood bitter-cress), *Agrostis stolonifera* (creeping bent-grass).

E. (Gleyed brown earth soils, woodland)
   Viola riviniana (common violet), *Primula vulgaris* (primrose), *Potentilla sterilis* (barren strawberry), *Deschampsia cespitosa* (tufted hair-grass), *Ajuga reptans* (bugle), *Chamaenerion angustifolium* (rosebay willow-herb), *Rubus idaeus* (raspberry), *Lysimachia nemorum* (yellow pimpernel), *Athyrium filix-femina* (lady-fern), *Equisetum sylvaticum* (wood horsetail).

F. (Acid brown earth soils, woodland)
   Stellaria holostea (greater stitchwort), *Oxalis acetosella* (wood-sorrel), *Holcus mollis* (creeping soft-grass), *Anemone nemorosa* (wood anemone), *Luzula sylvatica* (greater woodrush), *Lonicera periclymenum* (honeysuckle), *Dryopteris dilatata* (broad buckler-fern), *Dryopteris filix-mas* (male fern).

G. (Gley soils, woodland margin/open)
   Rumex acetosa (sorrel), *Holcus lanatus* (Yorkshire fog), *Ranunculus repens* (creeping buttercup), *Juncus effusus* (soft rush), *Cirsium palustre* (marsh thistle).

H. (Brown earth soils, open)
   Trifolium repens (white clover), *Plantago lanceolata* (ribwort), *Cerastium holosteoides* (common mouse-ear chickweed).

I. (Acid brown earth soils, open/woodland)
   Teucrium scorodonia (wood sage), *Pteridium aquilinum* (bracken), *Solidago virgaurea* (golden-rod), *Luzula pilosa* (hairy woodrush), *Blechnum spicant* (hard-fern), *Hypericum pulchrum* (slender St. John's wort), *Galium saxatile* (heath bedstraw), *Anthoxanthum odoratum* (sweet vernal-grass), *Digitalis purpurea* (foxglove), *Agrostis tenuis* (common bent-grass).

J. (Brown podzolic soils, open/woodland)
    Vaccinium myrtillus (bilberry), *Deschampsia flexuosa* (wavy hair-grass), *Succisa pratensis* (devil's-bit scabious), *Potentilla erecta* (common tormentil), *Luzula multiflora* (many-headed woodrush), *Festuca ovina* (sheep's fescue), *Agrostis canina* (brown bent-grass).
K. (Peaty podzolic soils, open/woodland)
    Molinia caerulea (purple moor-grass), *Erica cinerea* (bell-heather), *Calluna vulgaris* (ling).

1.2 Canopy and understorey species

1.2.1 **Constant trees** Two categories are provided: species occurring in more than 75% of the plots, of the original survey, are listed without brackets; those occurring in 20-75% of the plots are given in brackets. Thus, in PLOT TYPE 3 (page 43)

English elm (ash, field maple, oak)

indicates that English elm (ie *Ulmus procera)* occurred in at least 75% of the plots, and that ash, field maple and oak each occurred in at least 20% of the plots and not more than 74%.

1.2.2 **Constant saplings** These are treated as for constant trees, remembering that the breast height diameters of saplings are, by definition, less than 5 cm.

1.2.3 **Constant shrubs** These are treated as for constant trees, accepting that they are woody perennials that usually do not contribute to the canopy: they are members of the understorey, eg elder, hazel and holly.

1.2.4 **Trees (basal area)** This is an arbiter included to identify large trees. Where a species is listed, the basal area of the trees of that species within a 200 m^2 was at least 0·10 m^2.

2. Environment

2.1 Geographical distribution

Britain has been divided into 8 areas:

SW   south-west England
SE   south-east and south England
ME   midlands and East Anglia
NW   northern England, west of the Pennine watershed
NE   northern England, east of the Pennine watershed
Wa   north and south Wales
WS   west and south Scotland
ES   east Scotland

Regions which contained over 30% of the examples of a given plot type, as recorded in the original survey, are listed without brackets; regions within brackets have less than 30% of the examples of a given plot type and are ranked in order of frequency.

2.2 Solid geology

Geological information was obtained from the 10 inch (1 : 625 000) Ordnance Survey Geological Map. Each plot was assigned to a geological series. Geological series associated with more than 20% of the plots of a particular plot type are indicated without brackets; less frequent series are listed in brackets and in diminishing order of frequency.

To reduce numbers of geological series to a manageable size, some were amalgamated:

Code	Abbreviated description	Actual description
A	Calc clay	Calcareous clays and Oxford clays
B	K marl/Lias	Keuper marls, all Lias series, Kimmeridge clay
C	Wealden	Hastings beds, Oldhaven, London clay, Wealden
D	Devonian	Devonian series
E	Oolite/Chalk	Corallian, Cornbrash, Chalk and Southern oolites
F	Carb li/Mag li	Carboniferous and Magnesian limestone
G	Mill grit//Coal mea	Northern oolites, all Coal measures, Millstone grits
H	Silur/Ordov	Silurian and Ordovician series
I	Red s st	Red sandstone series and other sandstone
K	Ign/Metam	Residual igneous and metamorphic types

2.3 Altitude (m)

The mean altitude, in metres, of each replicate of a plot type was calculated from data on 2½ inch (1:25 000) Ordnance Survey maps. The average altitudes were divided into 3 zones:

(low)	(medium)	(high)
51-98 m	99-144 m	145-191 m

2.4 Slope (0°)

During the original field survey, the slope of the plots was measured in degrees using a Blume-Leiss clinometer. Three categories were recognised:

(low)	(medium)	(high)
3-10°	11-18°	19-26°

2.5 Rainfall (cm)

The average annual rainfall for each plot was taken from the Climatological Atlas of the British Isles (1952). As with altitude and slope, means were calculated and categorised as follows:

(low)	(medium)	(high)
64-93 cm	94-123 cm	124-152 cm

2.6 **Soil**

Soil samples from the top 10 cm were taken from the centre of each plot, and pH measurements were made with a glass electrode pH meter. Measurements were taken as soon after collection as possible, suspending soil in distilled water. Means were calculated and arranged in 3 groups:

(low)	(medium)	(high)
3·9-4·9	5·0-5·8	5·9-6·8

2.7 **LOI (percentage loss on ignition)**

LOI was determined from air-dried soil heated to 450° in a muffle furnace. Means were calculated and arranged in the following groups:

(low)	(medium)	(high)
10·0-28·0%	28·1-45·0%	45·1-63·0%

3. **General descriptions**

These have been prepared to complement the vegetational and environmental data and to stress a number of other features.

The plot types are considered to be infrequent, of average occurrence or frequent, if they account for less than 30, between 31 and 70, or more than 70 of the 1648 plots of the original survey (16 plots at each woodland location).

Heterogeneity can be measured within a plot or at the level of a relatively large tract of woodland. Sometimes, different plots within an area of woodland will belong to one plot type, whereas in other instances many types may occur within a comparable area. For example, in East Anglian woods, few plot types are represented (i.e. 2, 3 and 6), whereas a much greater range occurs in western Scotland (i.e. 18, 22, 25, 26, 29, 30, 31). The East Anglian woods are, therefore, more uniform than their Scottish equivalents. Some plot types have a restricted geographical distribution, eg type 3 which is associated with the calcareous boulder clay limited to East Anglia, whereas type 21, which occurs on a range of bedrocks, is present more or less throughout Britain.

Within plots, it was always possible to identify representatives of more than one species group, as described by Bunce (1977). Plots with representatives of 3·0-5·0, 5·1-7·0 and 7·1-9·0 species groups are rated as having low heterogeneity, medium heterogeneity and high heterogeneity respectively. Thus, PLOT TYPE 30 *SUCCISA PRATENSIS/HOLCUS MOLLIS* (DEVIL'S-BIT SCABIOUS/CREEPING SOFT-GRASS) is probably the most heterogeneous, and PLOT TYPE 17 *PTERIDIUM AQUILINUM/ RUBUS FRUTICOSUS* (BRACKEN/BRAMBLE) on Wealden clay is the least heterogeneous.

The number of species groups within a plot type gives an estimate of diversity, but this estimate is not always the same as that judged from the total number of species recorded from a plot type. The latter have been graded into 3 categories—low with 80-141 species, medium with 141-200 species, and high with more than 200 species.

The relationships between different plot types were deduced by ordination (Hill 1973). In the plot descriptions that follow, the 3 nearest plot types have been arranged in decreasing affinity. There may be few or many species within the 200 m² plots which may extensively cover the ground or 'occupy' very little of it. To obtain estimates of occupancy by herbs, means were calculated of eye assessments of percentage ground cover. Subsequently, the means were put into one of 3 categories—high, 81-100%; medium, 61-80%; low, 30-60%. Estimates were also made of the cover of litter, rock, bare ground, and ground bryophytes, but these are not mentioned in the descriptions unless the means were greater than 80%, 3%, 8% and 20%, respectively. The density of canopy was inferred from measurements of basal area, assuming that the tree canopy was dense if there was a basal area of $0.538-0.770$ m^2 per 200 m^2, average if between $0.305-0.537$ m^2 per 200 m^2, and open if between $0.071-0.304$ m^2 per 200 m^2. Frequency of shrub layer was described by 3 categories—usually, 46-60% of the plots; often, 31-45% of the plots; and rarely, 10-30% of the plots. The same ranges were used for describing occurrence of saplings.

The final paragraph of the general descriptions gives a succinct assessment of the different plot types, bringing together knowledge of the classification described in this booklet, and an appreciation of terms in common usage. In a few instances, the assessments concur with Tansley's (1947) descriptions, but it should be remembered that he was making generalisations about large areas which, probably, included more than one of the plot types identified within the present classification. The short descriptions are supported by outline descriptions of the characteristic soil groups.

Having evolved a set of plot types using a set of objective procedures, it would be worthwhile to compare them with the groupings produced by the Zürich-Montpellier approach for woodlands in northern Europe. Kiellund-Lund (1973) concentrated on the Vaccinio-Piceetia of Scandinavia, Hartman and Jahn (1967) on the Fagetea of Germany, and Durin et al. (1968) on the Quercetia robori-petreae of northern France. These associations are difficult to distinguish, and their names should only be used as guides. In cases of doubt, the classes proposed by Kiellund-Lund have been taken. Usually, British counterparts of European woodland assemblages contained more Atlantic species and greater numbers of species not usually associated with woodlands in continental Europe. The significance of these assemblages will be further illustrated when the second booklet is produced, showing how whole tracts of woodland can be assigned to a single class, taking account of the differing frequency of species recorded when random plots are surveyed.

ACKNOWLEDGEMENTS
M. W. Shaw was joint leader of the project and was instrumental in the design of the field procedure. He also carried out the computer analysis, and participated throughout in the development of the approach.

Many people have made valuable contributions at different stages of this project, and in particular the following permanent assistants at ITE Merlewood—C. J. Barr, Mrs. Wendy Bowen, Mrs. Carole Helliwell and Mrs. Judith Johnston. A. H. F. Brown, A. D. Horrill and J. M. Sykes, also at Merlewood, contributed valuable advice and assistance with field work in 1971, and M. O. Hill at ITE Bangor gave statistical advice and analysed some of the data in the Computing Laboratory of the University College of North Wales, Bangor, whose co-operation is also acknowledged. The following temporary staff also participated, largely in the collection of field data: N. E. Barber, P. A. Bassett, S. Daggitt, J. C. Holmes, I. P. Howes, T. J. Moss, G. K. A. Reynolds, P. L. Rye, Mrs. C. Sargent (now on the permanent staff of ITE), Miss C. Smith, D. J. Taylor and P. Wilkins. More recently, Prof. F. T. Last has guided the booklet through its final stages. The cover illustration was drawn by C. B. Benefield.

Our thanks also go to the many landowners and tenants who provided access to their woodlands, often offering hospitality and interest in the project.

REFERENCES

Allorge, P. 1922. Les associations végétales du Vexin Francais. *Revue gén. Bot.,* **33/34.**
Birse, E. L. & Robertson, J. S. 1976. *Plant communities and soils of the lowland and southern upland regions of Scotland.* Aberdeen: Macaulay Institute for Soil Research.
Braun-Blanquet, J. & De Leeuw, W. C. 1936. Vegetationskizze von Ameland. *Ned. kruidk. Archf,* **46,** 359-393.
Braun-Blanquet, J. & Tüxen, R. 1952. Irische Pflanzengesellschaften. Die Pflanzenwelt Irlands. Ergebnisse der 9 I.P.E. durch Irland 1949. *Veröff. geobot. Inst., Zürich,* **25,** 224-420.
Braun-Blanquet, J., Sissingh, G. & Vlieger, J. 1939. *Klasse der Vaccinio-Piceetea.* (Prodromus der Pflanzengesellschaften 6.) Montpellier.
Bunce, R. G. H. 1977. The range of variation in the pinewoods. In: *Native pinewoods of Scotland,* edited by R. G. H. Bunce and J. N. R. Jeffers, 11-25. Cambridge: Institute of Terrestrial Ecology.
Bunce, R. G. H. & Shaw, M. W. 1973. A standardised procedure for ecological survey. *J. environ. Manage.,* **1,** 239-258.
Cajander, A. K. 1921. Uber Waldtypen in allgemeinen I. *Acta for. fenn.,* **20,** 1-41.
Clapham, A. R., Tutin, T. G. & Warburg, E. F. 1962. *Flora of the British Isles.* 2nd ed. Cambridge: Cambridge University Press.
Durin, L., Géhu, J.-M., Noirfalise, A. & Sougnez, N. 1968. Les hêtraies atlantiques et leur essaim climacique dans le nordouest et l'ouest de la France. *Bull. Soc. bot. N. Fr.,* num. spèc. 20e anniv. 60-89.
Hartman, F. K. & Jahn, C. 1967. *Waldgesellschaften des mitteleuropäischen Gebirgsraumes nördlich der Alpen.* Stuttgart: Fischer.
Hill, M. O. 1973. Reciprocal averaging: an eigenvector method of ordination. *J. Ecol.,* **61,** 237-249.
Hill, M. O., Bunce, R. G. H. & Shaw, M. W. 1975. Indicator species analysis: a divisive polythetic method of classification and its application to a survey of native pinewoods in Scotland. *J. Ecol.,* **63,** 597-613.
Kiellund-Lund, J. 1967. Zur systematik der Kiefernwälder Fennoscandiens. *Mitt flor-soz. ArbGemein.* no. 11/12, 127-141.
Kiellund-Lund, J. 1973. A classification of Scandinavian forest vegetation for mapping purposes. *IBP i Norden,* **11,** 173-206.
Kleist, C. 1929. Récherches phytosociologiques sur les tourbières de la région des dures de la rive droite de la Vistule aux environs de Varsovie. *Bull. int. Acad. pol. Sci. Lett.* Sér. B, 41-104.
Klötzli, F. 1970. Eichen-, Edellaub und Bruchwälder der Britischen Inseln. *Schweiz. Z. Forstwes.,* **121,** 329-366.
Koch, W. 1926. Die Vegetationseinheiten der Linthebene. *Jb. St Gall. naturw, Ges.,* **61,** 219-225.
Libbert, W. 1933. Die Vegetationseinheiten der neumärkischen Staubecken Landschaft 2. *Verh. bot. Ver. Prov. Brandenb.,* **75,** 229-348.
McVean, D. N. & Ratcliffe, D.A. 1962. *Plant communities of the Scottish highlands: a study of Scottish mountain, moorland and forest vegetation.* London: H.M.S.O.
Passarge, H. 1956. *Die Wälder von Magdeburgerforth (N-W Flämig).* Thesis. Deutsche Akademie der Landwirtschaftswissenschaften.
Scamoni, A. & Passarge, H. 1959. Gedanken zu einer natürlichen Ordnung der Waldgesellschaften. *Arch. Forstw.,* **8,** 386-426.
Schwickerath, M. 1937. *III. Jber. Gruppe Preussen-Rheinl. dtsch. Forstver.* Berlin.
Seibert, P. 1969. Uber das Aceri-Fraxinetum als vikariierende Gesellschaft des Galio-Carpinetum am Rande der Bayerischen Alpen. *Vegetatio,* **17,** 165-175.
Tansley, A. G. 1947. *The British islands and their vegetation.* 2nd ed. Cambridge: Cambridge University Press.
Tüxen R. 1937. Die Pflanzengesellschaften Nordwestdeutschlands. *Mitt. flor-soz. ArbGemein.,* no. 3, 1-170.
Tüxen, R. 1951. Eindrücke während der Pflanzengeographischen Exkursion durch Süd-Schweden. *Vegetatio,* **3,** 149-172.

Tüxen, R. 1955. Das System der nordwestdeutschen Pflanzengesellschaften. *Mitt. flor-soz. ArbGemein.* NS. no. 5, 155-176.

Watson, E. V. 1955. *British mosses and liverworts.* Cambridge: Cambridge University Press.

THE KEY

STEP ONE

INDICATOR SPECIES

Negative   *Positive*

CIRCAEA LUTETIANA   Anthoxanthum odoratum
  (ENCHANTER'S NIGHTSHADE)   (sweet vernal-grass)
FRAXINUS EXCELSIOR   Deschampsia flexuosa
  (ASH)   (wavy hair-grass)
GEUM URBANUM   Galium saxatile
  (HERB BENNET)   (heath bedstraw)
MERCURIALIS PERENNIS   Pteridium aquilinum
  (DOG'S MERCURY)   (bracken)

EURHYNCHIUM PRAELONGUM   Polytrichum spp

Score −1 or less, go to STEP TWO
Score   0 or more, go to STEP SEVENTEEN

STEP TWO

INDICATOR SPECIES

Negative   *Positive*

ACER CAMPESTRE   Athyrium filix-femina
  (COMMON MAPLE)   (lady-fern)
ARUM MACULATUM   Dryopteris dilatata
  (LORDS-AND-LADIES)   (broad buckler-fern)
CORYLUS AVELLANA   Holcus mollis
  (HAZEL)   (creeping soft-grass)
MERCURIALIS PERENNIS   Lysimachia nemorum
  (DOG'S MERCURY)   (yellow pimpernel)
                                         Oxalis acetosella
                                           (wood-sorrel)

                                         Mnium hornum

Score −1 or less, go to STEP THREE
Score   0 or more, go to STEP TEN

INDICATOR SPECIES

Negative	Positive

ACER PSEUDOPLATANUS
(SYCAMORE)
SAMBUCUS NIGRA
(ELDER)
ULMUS PROCERA
(ENGLISH ELM)

Brachypodium sylvaticum
(slender false-brome)
Carex sylvatica
(wood sedge)
Corylus avellana
(hazel)
Crataegus monogyna
(common hawthorn)
Lonicera periclymenum
(honeysuckle)
Rubus fruticosus
(bramble)
Viola riviniana
(common violet)

Score 2 or less, go to STEP FOUR
Score 3 or more, go to STEP SEVEN

INDICATOR SPECIES

Negative	Positive

ENDYMION NON-SCRIPTUS
(BLUEBELL)
FAGUS SYLVATICA
(BEECH)
HEDERA HELIX
(IVY)
RUBUS FRUTICOSUS
(BRAMBLE)
SILENE DIOICA
(RED CAMPION)

Acer campestre
(common maple)
Circaea lutetiana
(enchanter's nightshade)
Ulmus procera
(English elm)

Fissidens taxifolius
Thamnium alopecurum

Score 0 or less, go to STEP FIVE
Score 1 or more, go to STEP SIX

INDICATOR SPECIES

Negative	*Positive*
CIRCAEA LUTETIANA (ENCHANTER'S NIGHTSHADE) CORYLUS AVELLANA (HAZEL) BRACHYTHECIUM RUTABULUM	Anthriscus sylvestris (cow parsley) Bromus ramosus (hairy brome) Crataegus monogyna (common hawthorn) Heracleum sphondylium (hogweed) Poa trivialis (rough meadow-grass) Silene dioica (red campion) Ulmus procera (English elm)

Score 1 or less,	*TYPE 1*	URTICA DIOICA/RUBUS FRUTICOSUS (STINGING NETTLE/BRAMBLE)
Score 2 or more,	*TYPE 2*	BROMUS RAMOSUS/MERCURIALIS PERENNIS (HAIRY BROME/DOG'S MERCURY)

INDICATOR SPECIES

Negative	*Positive*
ACER CAMPESTRE (COMMON MAPLE) AGROSTIS STOLONIFERA (CREEPING BENT-GRASS) ULMUS PROCERA (ENGLISH ELM) VIOLA RIVINIANA (COMMON VIOLET) EURHYNCHIUM PRAELONGUM THAMNIUM ALOPECURUM	Acer pseudoplatanus (sycamore) Dryopteris dilatata (broad buckler-fern) Hedera helix (ivy) Eurhynchium striatum

Score −1 or less,	*TYPE 3*	AGROSTIS STOLONIFERA/ MERCURIALIS PERENNIS (CREEPING BENT-GRASS/DOG'S MERCURY)
Score 0 or more,	*TYPE 4*	ARUM MACULATUM/MERCURIALIS PERENNIS (LORDS-AND-LADIES/DOG'S MERCURY)

INDICATOR SPECIES

Negative   *Positive*

BRACHYPODIUM SYLVATICUM   Ilex aquifolium
  (SLENDER FALSE-BROME)   (holly)
FILIPENDULA ULMARIA   Luzula pilosa
  (MEADOW-SWEET)   (hairy woodrush)
GALIUM APARINE   Mercurialis perennis
  (GOOSEGRASS)   (dog's mercury)
GERANIUM ROBERTIANUM
  (HERB ROBERT)
GLECHOMA HEDERACEA
  (GROUND IVY)
POA TRIVIALIS
  (ROUGH MEADOW-GRASS)
URTICA DIOICA
  (STINGING NETTLE)

Score −1 or less, go to STEP EIGHT
Score   0 or more, go to STEP NINE

INDICATOR SPECIES

Negative   *Positive*

EPILOBIUM MONTANUM   Arum maculatum
  (BROAD-LEAVED WILLOW-HERB)   (lords-and-ladies)
GLECHOMA HEDERACEA   Circaea lutetiana
  (GROUND IVY)   (enchanter's nightshade)
MERCURIALIS PERENNIS   Euonymus europaeus
  (DOG'S MERCURY)   (spindle)
SAMBUCUS NIGRA   Hedera helix
  (ELDER)   (ivy)
VIOLA RIVINIANA   Ligustrum vulgare

(COMMON VIOLET)   (common privet)

Score 0 or less,   *TYPE 5*   GLECHOMA HEDERACEA/
                                   MERCURIALIS PERENNIS
                                     (GROUND IVY/DOG'S MERCURY)
Score 1 or more,   *TYPE 6*   LISTERA OVATA/HEDERA HELIX
                                     (TWAYBLADE/IVY)

INDICATOR SPECIES

Negative	*Positive*
ACER CAMPESTRE	Sorbus aucuparia
(COMMON MAPLE)	(rowan)
DESCHAMPSIA CESPITOSA	Tamus communis
(TUFTED HAIR-GRASS)	(black bryony)
GALEOBDOLON LUTEUM	Taxus baccata
(YELLOW ARCHANGEL)	(yew)
HEDERA HELIX	Teucrium scorodonia
(IVY)	(wood sage)
EURHYNCHIUM PRAELONGUM	Hypnum cupressiforme

Score −1 or less,   *TYPE 7*   *CAREX SYLVATICA/*
                                            RUBUS FRUTICOSUS
                                            (WOOD SEDGE/BRAMBLE)

Score   0 or more,   *TYPE 8*   *MERCURIALIS PERENNIS/*
                                            RUBUS FRUTICOSUS
                                            (DOG'S MERCURY/BRAMBLE)

INDICATOR SPECIES

Negative	*Positive*
CORYLUS AVELLANA	Angelica sylvestris
(HAZEL)	(wild angelica)
LONICERA PERICLYMENUM	Chrysosplenium oppositifolium
(HONEYSUCKLE)	(opposite-leaved golden saxifrage)
QUERCUS SPP	Filipendula ulmaria
(OAK)	(meadow-sweet)
RUBUS FRUTICOSUS	Galium aparine
(BRAMBLE)	(goosegrass)
	Ranunculus repens
MNIUM HORNUM	(creeping buttercup).

Score −1 or less, go to STEP ELEVEN
Score   0 or more, go to STEP FOURTEEN

INDICATOR SPECIES

Negative	Positive
ATHYRIUM FILIX-FEMINA (LADY-FERN) *DIGITALIS PURPUREA* (FOXGLOVE) *DRYOPTERIS DILATATA* (BROAD BUCKLER-FERN)	*Brachypodium sylvaticum* (slender false-brome) *Carex sylvatica* (wood sedge) *Fragaria vesca* (wild strawberry) *Potentilla sterilis* (barren strawberry) *Primula vulgaris* (primrose) *Veronica chamaedrys* (germander speedwell) *Viola riviniana* (common violet)

Score 0 or less, go to STEP TWELVE
Score 1 or more, go to STEP THIRTEEN

INDICATOR SPECIES

Negative	Positive
DESCHAMPSIA CESPITOSA (TUFTED HAIR-GRASS) *FAGUS SYLVATICA* (BEECH) *PTERIDIUM AQUILINUM* (BRACKEN) *SORBUS AUCUPARIA* (ROWAN)	*Athyrium filix-femina* (lady-fern) *Cardamine flexuosa* (wood bitter-cress) *Corylus avellana* (hazel) *Veronica montana* (wood speedwell) *Eurhynchium striatum* *Thuidium tamariscinum*

Score 0 or less,	*TYPE 9*	*ENDYMION NON-SCRIPTUS/RUBUS FRUTICOSUS* (BLUEBELL/BRAMBLE)
Score 1 or more,	*TYPE 10*	*ATHYRIUM FILIX-FEMINA/RUBUS FRUTICOSUS* (LADY-FERN/BRAMBLE)

INDICATOR SPECIES

Negative	*Positive*

BETULA SPP
  (BIRCH)
POTENTILLA STERILIS
  (BARREN STRAWBERRY)

THUIDIUM TAMARISCINUM

Positive

Brachypodium sylvaticum
  (slender false-brome)
Deschampsia cespitosa
  (tufted hair-grass)
Geum urbanum
  (herb bennet)
Sanicula europaea
  (sanicle)
Stachys sylvatica
  (hedge woundwort)
Stellaria holostea
  (greater stitchwort)
Ulmus glabra
  (wych elm)

Score 0 or less,   *TYPE 11*   *POTENTILLA STERILIS/RUBUS FRUTICOSUS*
(BARREN STRAWBERRY/BRAMBLE)

Score 1 or more,   *TYPE 12*   *GEUM URBANUM/MERCURIALIS PERENNIS*
(HERB BENNET/DOG'S MERCURY)

INDICATOR SPECIES

Negative	*Positive*

ANGELICA SYLVESTRIS
  (WILD ANGELICA)
HERACLEUM SPHONDYLIUM
  (HOGWEED)
SILENE DIOICA
  (RED CAMPION)

Positive

Agrostis tenuis
  (common bent-grass)
Cirsium palustre
  (marsh thistle)
Fraxinus excelsior
  (ash)
Holcus lanatus
  (Yorkshire fog)
Prunella vulgaris
  (self-heal)
Trifolium repens
  (white clover)

Thuidium tamariscinum

Score 2 or less, go to STEP 15
Score 3 or more, go to STEP 16

INDICATOR SPECIES

Negative	Positive
ACER PSEUDOPLATANUS (SYCAMORE)	*Alnus glutinosa* (alder)
FRAXINUS EXCELSIOR (ASH)	*Betula* spp (birch)
VERONICA MONTANA (WOOD SPEEDWELL)	*Galium palustre* (marsh bedstraw)
	Lonicera periclymenum (honeysuckle)
	Salix spp (willow)
	Solanum dulcamara (woody nightshade)
	Valeriana officinalis (valerian)

Score 0 or less,	*TYPE 13*	*CHRYSOSPLENIUM OPPOSITIFOLIUM/ MERCURIALIS PERENNIS* (OPPOSITE-LEAVED GOLDEN SAXIFRAGE/DOG'S MERCURY)
Score 1 or more,	*TYPE 14*	*CHRYSOSPLENIUM OPPOSITIFOLIUM/ RUBUS FRUTICOSUS* (OPPOSITE-LEAVED GOLDEN SAXIFRAGE/BRAMBLE)

INDICATOR SPECIES

Negative	Positive
ARRHENATHERUM ELATIUS (OAT-GRASS)	*Brachypodium sylvaticum* (slender false-brome)
EPILOBIUM MONTANUM (BROAD-LEAVED WILLOW-HERB)	*Deschampsia cespitosa* (tufted hair-grass)
GALIUM PALUSTRE (MARSH BEDSTRAW)	*Galium saxatile* (heath bedstraw)
LOTUS PEDUNCULATUS (LARGE BIRDSFOOT TREFOIL)	*Rubus idaeus* (raspberry)
STELLARIA HOLOSTEA (GREATER STITCHWORT)	*Viola riviniana* (common violet)

Score 0 or less,	*TYPE 15*	*GALIUM PALUSTRE/AGROSTIS TENUIS* (MARSH BEDSTRAW/COMMON BENT-GRASS)
Score 1 or more,	*TYPE 16*	*CIRSIUM PALUSTRE/AGROSTIS TENUIS* (MARSH THISTLE/COMMON BENT-GRASS)

INDICATOR SPECIES

Negative	Positive

FAGUS SYLVATICA
  (BEECH)
LONICERA PERICLYMENUM
  (HONEYSUCKLE)
QUERCUS SPP
  (OAK)
RUBUS FRUTICOSUS
  (BRAMBLE)

Agrostis canina
  (brown bent-grass)
Anthoxanthum odoratum
  (sweet vernal-grass)
Galium saxatile
  (heath bedstraw)
Potentilla erecta
  (common tormentil)
Ranunculus repens
  (creeping buttercup)

Rhytidiadelphus squarrosus

Score −1 or less, go to STEP EIGHTEEN
Score   0 or more, go to STEP TWENTY-FIVE

INDICATOR SPECIES

Negative	Positive

CHAMAENERION
  ANGUSTIFOLIUM
  (ROSEBAY WILLOW-HERB)
DESCHAMPSIA FLEXUOSA
  (WAVY HAIR-GRASS)
HOLCUS LANATUS
  (YORKSHIRE FOG)

DICRANELLA HETEROMALLA

Athyrium filix-femina
  (lady-fern)
Corylus avellana
  (hazel)
Dryopteris filix-mas
  (male fern)
Holcus mollis
  (creeping soft-grass)
Oxalis acetosella
  (wood-sorrel)
Stellaria holostea
  (greater stitchwort)

Score 0 or less, go to STEP NINETEEN
Score 1 or more, go to STEP TWENTY-TWO

INDICATOR SPECIES

Negative	*Positive*
BLECHNUM SPICANT	Carpinus betulus
(HARD-FERN)	(hornbeam)
HEDERA HELIX	Chamaenerion angustifolium
(IVY)	(rosebay willow-herb)
ILEX AQUIFOLIUM	Holcus lanatus
(HOLLY)	(Yorkshire fog)
SORBUS AUCUPARIA	Juncus effusus
(ROWAN)	(soft rush)
DICRANUM SCOPARIUM	
MNIUM HORNUM	

Score −1 or less, go to STEP TWENTY
Score   0 or more, go to STEP TWENTY-ONE

INDICATOR SPECIES

Negative	*Positive*
FAGUS SYLVATICA	Deschampsia flexuosa
(BEECH)	(wavy hair-grass)
ILEX AQUIFOLIUM	Lonicera periclymenum
(HOLLY)	(honeysuckle)
	Vaccinium myrtillus
ISOPTERYGIUM ELEGANS	(bilberry)
	Dicranum scoparium
	Hypnum cupressiforme
	Leucobryum glaucum
	Polytrichum spp

Score 0 or less,	TYPE 17	PTERIDIUM AQUILINUM/RUBUS FRUTICOSUS (BRACKEN/BRAMBLE)
Score 1 or more,	TYPE 18	DESCHAMPSIA FLEXUOSA/PTERIDIUM AQUILINUM (WAVY HAIR-GRASS/BRACKEN)

INDICATOR SPECIES

Negative	*Positive*

*CHAMAENERION
   ANGUSTIFOLIUM*
  (ROSEBAY WILLOW-HERB)
CIRSIUM PALUSTRE
  (MARSH THISTLE)
DRYOPTERIS DILATATA
  (BROAD BUCKLER-FERN)
DRYOPTERIS FILIX-MAS
  (MALE FERN)

Carpinus betulus
  (hornbeam)
Fagus sylvatica
  (beech)
Poa trivialis
  (rough meadow-grass)
Quercus spp
  (oak)

Mnium hornum
Polytrichum spp

Score 0 or less,   *TYPE 19*   *CHAMAENERION ANGUSTIFOLIUM/
PTERIDIUM AQUILINUM*
  (ROSEBAY WILLOW-HERB/
  BRACKEN)

Score 1 or more,   *TYPE 20*   *CHAMAENERION ANGUSTIFOLIUM/
RUBUS FRUTICOSUS*
  (ROSEBAY WILLOW-HERB/BRAMBLE)

INDICATOR SPECIES

Negative	*Positive*

ACER PSEUDOPLATANUS
  (SYCAMORE)
ATHYRIUM FILIX-FEMINA
  (LADY-FERN)
DIGITALIS PURPUREA
  (FOXGLOVE)
DRYOPTERIS DILATATA
  (BROAD BUCKLER-FERN)
FRAXINUS EXCELSIOR
  (ASH)
SORBUS AUCUPARIA
  (ROWAN)

Holcus mollis
  (creeping soft-grass)
Luzula pilosa
  (hairy woodrush)
Pteridium aquilinum
  (bracken)
Stellaria holostea
  (greater stitchwort)

Score −1 or less, go to STEP TWENTY-THREE
Score   0 or more, go to STEP TWENTY-FOUR

INDICATOR SPECIES

Negative	Positive

ENDYMION NON-SCRIPTUS
  (BLUEBELL)
GALIUM SAXATILE
  (HEATH BEDSTRAW)
PTERIDIUM AQUILINUM
  (BRACKEN)

Betula spp
  (birch)
Blechnum spicant
  (hard-fern)
Corylus avellana
  (hazel)
Fraxinus excelsior
  (ash)
Hedera helix
  (ivy)
Luzula sylvatica
  (greater woodrush)

Thuidium tamariscinum

Score −1 or less,	*TYPE 21*	*OXALIS ACETOSELLA/PTERIDIUM AQUILINUM* (WOOD-SORREL/BRACKEN)
Score 0 or more,	*TYPE 22*	*BLECHNUM SPICANT/RUBUS FRUTICOSUS* (HARD-FERN/BRAMBLE)

INDICATOR SPECIES

Negative	Positive

ACER PSEUDOPLATANUS
  (SYCAMORE)
DRYOPTERIS DILATATA
  (BROAD BUCKLER-FERN)
DRYOPTERIS FILIX-MAS
  (MALE FERN)
ENDYMION NON-SCRIPTUS
  (BLUEBELL)
OXALIS ACETOSELLA
  (WOOD-SORREL)

Corylus avellana
  (hazel)
Crataegus monogyna
  (common hawthorn)
Festuca gigantea
  (tall brome)
Galium saxatile
  (heath bedstraw)
Viola riviniana
  (common violet)

Score 0 or less,	*TYPE 23*	*HOLCUS MOLLIS/PTERIDIUM AQUILINUM* (CREEPING SOFT-GRASS/BRACKEN)
Score 1 or more,	*TYPE 24*	*LUZULA PILOSA/PTERIDIUM AQUILINUM* (HAIRY WOODRUSH/BRACKEN)

INDICATOR SPECIES

Negative	Positive

Negative

CALLUNA VULGARIS
 (LING)
DESCHAMPSIA FLEXUOSA
 (WAVY HAIR-GRASS)
SORBUS AUCUPARIA
 (ROWAN)
VACCINIUM MYRTILLUS
 (BILBERRY)

Positive

Athyrium filix-femina
 (lady-fern)
Epilobium montanum
 (broad-leaved willow-herb)
Lysimachia nemorum
 (yellow pimpernel)
Prunella vulgaris
 (self-heal)
Ranunculus repens
 (creeping buttercup)
Veronica chamaedrys
 (germander speedwell)

Score 0 or less, go to STEP TWENTY-SIX
Score 1 or more, go to STEP TWENTY-NINE

INDICATOR SPECIES

Negative

DRYOPTERIS DILATATA
 (BROAD BUCKLER-FERN)
OXALIS ACETOSELLA
 (WOOD-SORREL)
QUERCUS SPP
 (OAK)
RUBUS FRUTICOSUS
 (BRAMBLE)

MNIUM HORNUM

Positive

Calluna vulgaris
 (ling)
Erica cinerea
 (bell-heather)
Molinia caerulea
 (purple moor-grass)

Pleurozium schreberi
Sphagnum spp

Score −1 or less, go to STEP TWENTY-SEVEN
Score 0 or more, go to STEP TWENTY-EIGHT

INDICATOR SPECIES

Negative		*Positive*

QUERCUS SPP
(OAK)

Deschampsia cespitosa
  (tufted hair-grass)
Holcus mollis
  (creeping soft-grass)
Hypericum pulchrum
  (slender St. John's wort)
Lysimachia nemorum
  (yellow pimpernel)
Rubus idaeus
  (raspberry)
Viola riviniana
  (common violet)
Veronica chamaedrys
  (germander speedwell)

Eurhynchium praelongum
Thuidium tamariscinum

Score 1 or less,   *TYPE 25*   *GALIUM SAXATILE/DESCHAMPSIA FLEXUOSA*
(HEATH BEDSTRAW/WAVY HAIR-GRASS)

Score 2 or more,   *TYPE 26*   *POTENTILLA ERECTA/HOLCUS MOLLIS*
(COMMON TORMENTIL/ CREEPING SOFT-GRASS)

INDICATOR SPECIES

Negative	*Positive*

AGROSTIS TENUIS
  (COMMON BENT-GRASS)
GALIUM SAXATILE
  (HEATH BEDSTRAW)
OXALIS ACETOSELLA
  (WOOD-SORREL)

Agrostis canina
  (brown bent-grass)
Carex echinata
  (star sedge)
Carex nigra
  (common sedge)
Molinia caerulea
  (purple moor-grass)
Narthecium ossifragum
  (bog asphodel)
Potentilla erecta
  (common tormentil)

Sphagnum spp

Score 3 or less,   *TYPE 27*   *CALLUNA VULGARIS/PTERIDIUM AQUILINUM*
  (LING/BRACKEN)

Score 4 or more,   *TYPE 28*   *NARTHECIUM OSSIFRAGUM/MOLINIA CAERULEA*
  (BOG ASPHODEL/PURPLE MOOR-GRASS)

INDICATOR SPECIES

Negative	*Positive*

DRYOPTERIS DILATATA
  (BROAD BUCKLER-FERN)
	Cerastium holosteoides
	  (common mouse-ear chickweed)

DRYOPTERIS FILIX-MAS
  (MALE FERN)
	Plantago lanceolata
	  (ribwort)

HOLCUS MOLLIS
  (CREEPING SOFT-GRASS)
	Taraxacum spp
	  (dandelion)

OXALIS ACETOSELLA
  (WOOD-SORREL)
	Trifolium repens
	  (white clover)

VIOLA RIVINIANA (COMMON VIOLET)	

POLYTRICHUM SPP

Score −1 or less, go to STEP THIRTY
Score   0 or more, go to STEP THIRTY-ONE

INDICATOR SPECIES

Negative	*Positive*

AGROSTIS STOLONIFERA
  (CREEPING BENT-GRASS)
	Angelica sylvestris
	  (wild angelica)

QUERCUS SPP
  (OAK)
	Cynosurus cristatus
	  (crested dog's-tail)

RUBUS FRUTICOSUS
  (BRAMBLE)
	Epilobium montanum
	  (broad-leaved willow-herb)

MNIUM HORNUM	Primula vulgaris (primrose)
	Ranunculus repens (creeping buttercup)
	Succisa pratensis (devil's-bit scabious)

Score 0 or less,	TYPE 29	ANTHOXANTHUM ODORATUM/
		AGROSTIS TENUIS
		  (SWEET VERNAL-GRASS/
		  COMMON BENT-GRASS)

Score 1 or more,	TYPE 30	SUCCISA PRATENSIS/HOLCUS MOLLIS
		  (DEVIL'S-BIT SCABIOUS/
		  CREEPING SOFT-GRASS)

INDICATOR SPECIES

Negative	Positive

CERASTIUM HOLOSTEOIDES
  (COMMON MOUSE-EAR
  CHICKWEED)
LEONTODON SPP
  (HAWKBIT)
TRIFOLIUM REPENS
  (WHITE CLOVER)

Alnus glutinosa
  (alder)
Circaea lutetiana
  (enchanter's nightshade)
Glyceria fluitans
  (flote-grass)
Mentha aquatica
  (water mint)
Potentilla anserina
  (silverweed)
Pulicaria dysenterica
  (fleabane)
Salix spp
  (willow)

Score 0 or less,   *TYPE 31*

Score 1 or more,   *TYPE 32*

*TRIFOLIUM PRATENSE/HOLCUS
LANATUS*
  (RED CLOVER/YORKSHIRE FOG)
*EPILOBIUM PALUSTRE/CAREX
VESICARIA*
  (MARSH WILLOW-HERB/BLADDER
  SEDGE)

DETAILED DESCRIPTIONS OF TYPES OF WOODLAND VEGETATION
including distribution maps and colour photographs

The photographs were taken within random plots drawn from the plots within the relevant type. The pictures were taken as near as possible to the relocated centres and a check was made that the appropriate indicators were present. Not all these species appear in the photographs because of the restricted field of vision. The species listed in the captions are those which appear prominent and are therefore mainly major cover species, rather than the frequently less obvious indicators. The importance of the indicators is emphasised by this approach because of the high degree of similarity of the cover species in very dissimilar ecological conditions, a consequence of the limited number of dominants in ground vegetation in British woodlands. The captions are restricted to vegetation only to emphasise the botanical basis of the key. Site names are not used in order to avoid identification of a type with a particular place—the emphasis is on the nature of the species composition, rather than the more usual association with sites.

In general, the photographs emphasise the continuous nature of woodland vegetation and the high degree of similarity between adjacent types—the original basis for the development of numerical rules for their separation. The types are arbitrary points on the continuum, and the photographs are designed to provide an overall impression of the types, together with an example of the associated woodland structure. However, structure is not a determining factor in the classification and, as with other factors such as soil, is consistent with the types because of its association with the ground vegetation.

Finally, the photographs demonstrate that plots are based on actual samples and must, therefore contain a degree of variation within them. It is their overall composition which is important, rather than their dependence on single species, and the photographs are intended to supplement the numerical data in the detailed descriptions.

(All photographs were supplied by R. G. H. Bunce)

PLOT TYPE 1

URTICA DIOICA/RUBUS FRUTICOSUS (STINGING NETTLE/BRAMBLE) TYPE

VEGETATION

Key species
Constant species: *Rubus fruticosus* (bramble)

Plot dominants: *Rubus fruticosus* (bramble), *Mercurialis perennis* (dog's mercury), *Hedera helix* (ivy)

Selective species: *Sambucus nigra* (elder), *Urtica dioica* (stinging nettle), *Aesculus hippocastanum* (horse chestnut), *Mercurialis perennis* (dog's mercury), *Euonymus europaeus* (spindle), *Galium aparine* (goosegrass)

Species groups: A, B (D, E, F)

Canopy and understorey species

Constant trees	*Constant saplings*
(ash, beech, oak, sycamore)	(ash, sycamore)
Constant shrubs	*Trees (basal area)*
(hazel, elder)	oak, beech

ENVIRONMENT

Geographical distribution
SE, SW (Wa, ME, NE, NW)

Solid geology
Oolite/Chalk (B, C, A, F, G, H)

Altitude (m)	*Slope (°)*	*Rainfall (cm)*	*Soil pH*	*LOI*
99 *(med)*	8 *(low)*	81 *(low)*	5·9 *(high)*	15·3 *(low)*

GENERAL DESCRIPTION

A type of average occurrence in an average range of site types, with low heterogeneity and a low species complement, most closely related to types 2, 7 and 5. There is a medium ground cover, with a high cover of litter. The canopy is usually dense with a few saplings, and there is often an understorey.

This type would probably be referred to as mixed deciduous woodland on rather base-rich soils in lowland Britain. The soils are mainly rather eutrophic brown earths that are quite base-rich. The comparable phytosociological associations are probably Ulmo-Fraxinetum E Sjogren ap KL 1973 (Ulmo-Quercetum Tx 1951) and Querco-Fraxinetum Klötzli 1970.

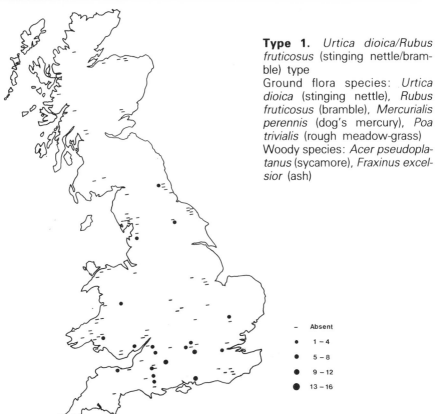

Type 1. *Urtica dioica/Rubus fruticosus* (stinging nettle/bramble) type
Ground flora species: *Urtica dioica* (stinging nettle), *Rubus fruticosus* (bramble), *Mercurialis perennis* (dog's mercury), *Poa trivialis* (rough meadow-grass)
Woody species: *Acer pseudoplatanus* (sycamore), *Fraxinus excelsior* (ash)

–	Absent
•	1 – 4
•	5 – 8
●	9 – 12
●	13 – 16

PLOT TYPE 2

BROMUS RAMOSUS/MERCURIALIS PERENNIS (HAIRY BROME/DOG'S MER-CURY) TYPE

VEGETATION

Key species

Constant species: *Poa trivialis* (rough meadow-grass), *Urtica dioica* (stinging nettle), *Silene dioica* (red campion), *Fraxinus excelsior* (ash), *Mercurialis perennis* (dog's mercury), *Rubus fruticosus* (bramble), *Hedera helix* (ivy), *Eurhynchium praelongum*

Plot dominants: *Rubus fruticosus* (bramble), *Mercurialis perennis* (dog's mercury), *Galeobdolon luteum* (yellow archangel), *Hedera helix* (ivy)

Selective species: *Anthriscus sylvestris* (cow parsley), *Bromus ramosus* (hairy brome), *Ulmus procera* (English elm), *Campanula trachelium* (bats-in-the-belfry), *Silene dioica* (red campion), *Milium effusum* (wood millet).

Species groups: A, B (D)

Canopy and understorey species

| *Constant trees* | *Constant saplings* |
| (ash, English elm, oak, sycamore, hawthorn, field maple) | English elm, ash, sycamore, hawthorn |

| *Constant shrubs* | *Trees (basal area)* |
| (elder) | (oak, ash, English elm) |

ENVIRONMENT

Geographical distribution
ME (NE, SE, SW)

Solid geology
K marl/Lias, Oolite/Chalk (G, C, A, F)

Altitude (m)	Slope (°)	Rainfall (cm)	Soil pH	LOI
71 *(low)*	14 *(med)*	74 *(low)*	5·5 *(med)*	12·6 *(low)*

GENERAL DESCRIPTION

A type of average occurrence in an average range of site types, with low heterogeneity and a low species complement, most closely related to types, 1, 7 and 4. There is usually a low ground cover with much bare ground. The canopy is usually dense, beneath which there is often an understorey.

This type would probably be described as a pedunculate oak/ash woodland growing under moist base-rich conditions. The soils are mainly brown earths, although often with some gleying. The comparable phytosociological association is probably Ulmo-Fraxinetum E Sjogren ap KL 1973 (Ulmo-Quercetum Tx 1951).

Type 2. *Bromus ramosus/Mercurialis perennis* (hairy brome/dog's mercury) type
Ground flora species: *Urtica dioica* (stinging nettle), *Rubus fruticosus* (bramble)
Woody species: *Quercus* spp (oak), *Rubus fruticosus* (bramble)

-   Absent
•   1 – 4
•   5 – 8
●   9 – 12
●   13 – 16

PLOT TYPE 3

AGROSTIS STOLONIFERA/MERCURIALIS PERENNIS (CREEPING BENT-GRASS/ DOG'S MERCURY) TYPE

VEGETATION

Key species

Constant species: *Fraxinus excelsior* (ash), *Mercurialis perennis* (dog's mercury), *Ulmus procera* (English elm), *Eurhynchium praelongum*

Plot dominants: *Mercurialis perennis* (dog's mercury), *Agrostis stolonifera* (creeping bent-grass)

Selective species: *Ulmus procera* (English elm), *Sambucus nigra* (elder), *Acer campestre* (common maple), *Mercurialis perennis* (dog's mercury), *Agrostis stolonifera* (creeping bent-grass), Thamnium alopecurum

Species groups: A, B (D, E)

Canopy and understorey species

Constant trees	*Constant saplings*
English elm (ash, field maple, oak)	(English elm)
Constant shrubs	*Trees (basal area)*
(hazel, elder)	English elm, ash

ENVIRONMENT

Geographical distribution	*Solid geology*
ME (NE, SE)	Calc clay (C, G)

Altitude (m)	Slope (°)	Rainfall (cm)	Soil pH	LOI
79 *(low)*	5 *(low)*	64 *(low)*	6·6 *(high)*	15·0 *(low)*

GENERAL DESCRIPTION

A type of average occurrence but in a restricted range of site types, with very low heterogeneity and a low species complement, most closely related to types 6, 4 and 7. There is a high ground cover and also much bare ground. The canopy is usually dense, with an understorey and with few saplings present.

This type would probably be referred to as English elm/ash woodland growing on calcareous clay. The soil is invariably a calcareous gley. The comparable phytosociological association is probably Ulmo-Fraxinetum E Sjogren ap KL 1973 (Ulmo-Quercetum Tx 1951).

Type 3. *Agrostis stolonifera/ Mercurialis perennis* (creeping bent-grass/dog's mercury) type
Ground flora species: *Glechoma hederacea* (ground ivy), *Rubus fruticosus* (bramble)
Woody species: *Ulmus procera* (English elm), *Sambucus nigra* (elder)

–	Absent
•	1 – 4
•	5 – 8
●	9 – 12
●	13 – 16

PLOT TYPE 4

ARUM MACULATUM/MERCURIALIS PERENNIS (LORDS-AND-LADIES/DOG'S MERCURY) TYPE

VEGETATION

Key species

Constant species: *Mercurialis perennis* (dog's mercury), *Fraxinus excelsior* (ash), *Circaea lutetiana* (enchanter's nightshade), *Rubus fruticosus* (bramble)

Plot dominants: *Mercurialis perennis* (dog's mercury), *Rubus fruticosus* (bramble), *Hedera helix* (ivy)

Selective species: *Iris foetidissima* (stinking iris), *Ligustrum vulgare* (common privet), *Arum maculatum* (lords-and-ladies), *Mercurialis perennis* (dog's mercury), *Sambucus nigra* (elder), *Fissidens taxifolius*

Species groups: A, B (E, F)

Canopy and understorey species

Constant trees	*Constant saplings*
(ash, sycamore, oak)	(ash, sycamore)

Constant shrubs	*Trees (basal area)*
(hazel)	ash (oak)

ENVIRONMENT

Geographical distribution
SW (SE, ME, NE, MW)

Solid geology
Oolite/Chalk, K marl/Lias (A, F, C)

Altitude (m)	*Slope (°)*	*Rainfall (cm)*	*Soil pH*	*LOI*
94 *(low)*	9 *(low)*	84 *(low)*	6·8 *(high)*	15·1 *(low)*

GENERAL DESCRIPTION

A type of infrequent occurrence in an average range of site types, with low heterogeneity and a low species complement, most closely related to types 7, 1 and 3. There is a high ground cover, but with bare ground and bryophytes also having a high cover. The canopy is usually dense with an understorey often present, as well as a low density of saplings.

This type would usually be called ash woodland on lowland, calcareous soils. The soil is usually a calcareous gley, although transitions are frequently present with brown earths. The comparable phytosociological associations are probably Ulmo-Fraxinetum E Sjogren ap KL 1973 (Ulmo-Quercetum Tx 1951) and Phyllitido-Fraxinetum Klötzli 1970.

Type 4. *Arum maculatum/Mercurialis perennis* (lords-and-ladies/dog's mercury) type
Ground flora species: *Mercurialis perennis* (dog's mercury), *Rubus fruticosus* (bramble)
Woody species: *Quercus* spp (oak), *Populus* spp (poplar), *Corylus avellana* (hazel)

–	Absent
•	1 – 4
•	5 – 8
●	9 – 12
●	13 – 16

PLOT TYPE 5

GLECHOMA HEDERACEA/MERCURIALIS PERENNIS (GROUND IVY/DOG'S MER-CURY) TYPE

VEGETATION

Key species

Constant species: *Corylus avellana* (hazel), *Rubus fruticosus* (bramble), *Fraxinus excelsior* (ash), *Geum urbanum* (herb bennet), *Crataegus monogyna* (common hawthorn), *Poa trivialis* (rough meadow-grass), *Urtica dioica* (stinging nettle), *Mercurialis perennis* (dog's mercury), *Viola riviniana* (common violet), *Eurhynchium praelongum*

Plot dominants: *Mercurialis perennis* (dog's mercury), *Rubus fruticosus* (bramble)

Selective species: *Glechoma hederacea* (ground ivy), *Geum urbanum* (herb bennet), *Brachypodium sylvaticum* (slender false-brome), *Urtica dioica* (stinging nettle), *Poa trivialis* (rough meadow-grass), *Acer campestre* (common maple).

Species groups: E, B (F, D, G, C)

Canopy and understorey species

Constant trees	Constant saplings
(ash, hawthorn, oak, field maple, birch)	(hawthorn, ash)

Constant shrubs	Trees (basal area)
hazel (elder)	ash (oak)

ENVIRONMENT

Geographical distribution	Solid geology
SW, ME (NE, SE, Wa, NW)	Oolite/Chalk, Calc clay (F, B, C, G, D)

Altitude (m)	Slope (°)	Rainfall (cm)	Soil pH	LOI
108 (med)	7 (low)	79 (low)	6·2 (high)	13·9 (low)

GENERAL DESCRIPTION

A type of frequent occurrence in an average range of site types, with medium heterogeneity and an average species complement, most closely related to types 7, 1 and 2. There is usually an average ground cover, but with bare ground and bryophytes also having much cover. The canopy is commonly of medium density, with an understorey usually present and with saplings in low density.

This type would probably be called mixed deciduous woodland. The soils are mainly calcareous brown earths, with gleying sometimes present. The comparable phytosociological associations are probably Ulmo-Fraxinetum E Sjogren ap KL 1973 (Ulmo-Quercetum Tx 1951) and Querco-Fraxinetum Klötzli 1970.

Type 5. *Glechoma hederacea/ Mercurialis perennis* (ground ivy/ dog's mercury) type
Ground flora species: *Mercurialis perennis* (dog's mercury), *Rubus fruticosus* (bramble), *Dryopteris filix-mas* (male fern), *Athyrium filix-femina* (lady-fern)
Woody species: *Fraxinus excelsior* (ash), *Corylus avellana* (hazel)

–	Absent
•	1 – 4
●	5 – 8
●	9 – 12
●	13 – 16

PLOT TYPE 6

LISTERA OVATA/HEDERA HELIX (TWAYBLADE/IVY) TYPE

VEGETATION

Key species

Constant species: *Fraxinus excelsior* (ash), *Galium aparine* (goosegrass), *Circaea lutetiana* (enchanter's nightshade), *Rubus fruticosus* (bramble), *Corylus avellana* (hazel), *Ligustrum vulgare* (common privet), *Hedera helix* (ivy), *Crataegus monogyna* (common hawthorn), *Urtica dioica* (stinging nettle)

Plot dominants: *Hedera helix* (ivy), *Mercurialis perennis* (dog's mercury), *Circaea lutetiana* (enchanter's nightshade), *Rubus fruticosus* (bramble)

Selective species: *Ligustrum vulgare* (common privet), *Listera ovata* (twayblade), *Ulmus carpinifolia* (smooth elm), *Iris foetidissima* (stinking iris), *Viola odorata* (sweet violet), *Galium aparine* (goosegrass)

Species groups: A, B (E, F, D)

Canopy and understorey species

Constant trees	Constant saplings
ash (oak, wych elm, hawthorn, field maple, sycamore)	(ash, hawthorn, English elm)

Constant shrubs	Trees (basal area)
hazel (privet)	oak (ash)

ENVIRONMENT

Geographical distribution	Solid geology
SW (ME, SE)	K marl/Lias, Oolite/Chalk (A, C)

Altitude (m)	Slope (°)	Rainfall (cm)	Soil pH	LOI
65 *(low)*	4 *(low)*	86 *(low)*	6·8 *(high)*	16·0 *(low)*

GENERAL DESCRIPTION

A type of average occurrence but in a restricted range of site types, with low heterogeneity and an average species complement, most closely related to types 3, 4 and 7. There is an average ground cover but with bare ground and bryophytes also having a high cover. There is a dense canopy with an understorey of hazel. Saplings are usually present in low density.

This type would probably be termed ash woodland mixed with pedunculate oak in lowland Britain. The soils are predominantly calcareous brown earths. The comparable phytosociological associations are probably Ulmo-Fraxinetum E Sjogren ap KL 1973 (Ulmo-Quercetum Tx 1951) and Querco-Fraxinetum Klötzli 1970.

Type 6. *Listera ovata/Hedera helix (twayblade/ivy) type*
Ground flora species: Mercurialis perennis (dog's mercury), *Hedera helix* (ivy), *Rubus fruticosus* (bramble), *Dryopteris filix-mas* (male fern)
Woody species: *Fraxinus excelsior* (ash), *Corylus avellana* (hazel), *Acer pseudoplatanus* (sycamore)

–	Absent
•	1 – 4
•	5 – 8
●	9 – 12
●	13 – 16

51

PLOT TYPE 7

CAREX SYLVATICA/RUBUS FRUTICOSUS (WOOD SEDGE/BRAMBLE) TYPE

VEGETATION

Key species

Constant species: *Fraxinus excelsior* (ash), *Rubus fruticosus* (bramble), *Corylus avellana* (hazel), *Viola riviniana* (common violet), *Eurhynchium praelongum*

Plot dominants: *Rubus fruticosus* (bramble), *Mercurialis perennis* (dog's mercury)

Selective species: *Acer campestre* (common maple), *Sorbus torminalis* (wild service tree), *Euphorbia amygdaloides* (wood spurge), *Corylus avellana* (hazel), *Carex sylvatica* (wood sedge), *Eurhynchium striatum*

Species groups: A, E, B (F, I)

Canopy and understorey species

Constant trees	*Constant saplings*
ash (oak, hawthorn, birch, field maple)	(ash, hawthorn, field maple)
Constant shrubs	*Trees (basal area)*
hazel (dogwood)	oak, ash

ENVIRONMENT

Geographical distribution   *Solid geology*
SW (ME, NE, SE, Wa)   K marl/Lias, Oolite/Chalk (A, D, F, C)

Altitude (m)	Slope (°)	Rainfall (cm)	Soil pH	LOI
100 *(med)*	10 *(low)*	84 *(low)*	5·8 *(med)*	13·9 *(low)*

GENERAL DESCRIPTION

A type of average occurrence but in a wide range of site types, with medium heterogeneity and an average species complement, most closely related to types 4, 5 and 1. There is usually an average ground cover with much bare ground and bryophytes also. There is a medium density canopy with a dense understorey and a medium density of saplings.

This type would usually be called mixed deciduous woodland, although ash would often be the main species together with pedunculate oak. The soil is usually a brown earth, although there is a tendency for gleying to take place. The comparable phytosociological associations are probably Ulmo-Fraxinetum E Sjogren ap KL 1973 (Ulmo-Quercetum Tx 1951) and Querco-Fraxinetum Klötzli 1970 or Corylo-Fraxinetum Br-Bl et Tx 1952.

Type 7. *Carex sylvatica/Rubus fruticosus* (wood sedge/bramble) type
Ground flora species: *Rubus fruticosus* (bramble), *Mercurialis perennis* (dog's mercury), *Dryopteris filix-mas* (male fern)
Woody species: *Fraxinus excelsior* (ash), *Corylus avellana* (hazel)

–	Absent
•	1 – 4
•	5 – 8
•	9 – 12
●	13 – 16

PLOT TYPE 8

MERCURIALIS PERENNIS/RUBUS FRUTICOSUS (DOG'S MERCURY/BRAMBLE)
TYPE

VEGETATION

Key species

Constant species:

Fraxinus excelsior (ash), *Mercurialis perennis* (dog's mercury), *Rubus fruticosus* (bramble), *Corylus avellana* (hazel), *Viola riviniana* (common violet), *Crataegus monogyna* (common hawthorn)

Plot dominants:

Rubus fruticosus (bramble), *Mercurialis perennis* (dog's mercury)

Selective species:

Taxus baccata (yew), *Carex flacca* (carnation-grass), *Convallaria majalis* (lily-of-the-valley), *Mercurialis perennis* (dog's mercury), *Fragaria vesca* (wild strawberry), *Rubus saxatilis* (stone bramble)

Species groups:

B, A (E, I, F, D)

Canopy and understorey species

Constant trees
(ash, sycamore, oak, yew, hawthorn)

Constant saplings
(ash, sycamore, hawthorn)

Constant shrubs
hazel

Trees (basal area)
(ash, oak, beech, yew)

ENVIRONMENT

Geographical distribution
NW (SE, NE, Wa, SW)

Solid geology
Carb li/Mag li (C, E)

Altitude (m)	Slope (°)	Rainfall (cm)	Soil pH	LOI
103 (med)	14 (med)	115 (med)	6·4 (high)	31·8 (med)

GENERAL DESCRIPTION

A type of infrequent occurrence in a restricted range of site types, with low heterogeneity and a low species complement, most closely related to types 7, 11 and 12. There is usually a low ground cover with a high proportion of rock. The canopy is quite dense and a dense understorey with an average density of saplings is present.

This type would usually be called ash woodland on limestone, although oak is often present. The soils are mainly brown earths, although there are some rendzinas. The comparable phytosociological associations are probably Ulmo-Fraxinetum E Sjogren ap KL 1973 (Ulmo-Quercetum Tx 1951) and Phyllitido-Fraxinetum Klötzli 1970 or Corylo-Fraxinetum Br-Bl et Tx 1952.

Type 8. *Mercurialis perennis/ Rubus fruticosus* (dog's mercury/ bramble) type
Ground flora species: *Mercurialis perennis* (dog's mercury), *Galium odoratum* (sweet woodruff)
Woody species: *Fraxinus excelsior* (ash), *Corylus avellana* (hazel)

–	Absent
•	1 – 4
•	5 – 8
●	9 – 12
●	13 – 16

PLOT TYPE 9

ENDYMION NON-SCRIPTUS/RUBUS FRUTICOSUS (BLUEBELL/BRAMBLE) TYPE

VEGETATION

Key species

Constant species: *Rubus fruticosus* (bramble), *Fraxinus excelsior* (ash), *Dryopteris filix-mas* (male fern), *Dryopteris dilatata* (broad buckler-fern), *Acer pseudoplatanus* (sycamore), *Oxalis acetosella* (wood-sorrel), *Eurhynchium praelongum*

Plot dominants: *Rubus fruticosus* (bramble), *Pteridium aquilinum* (bracken), *Dryopteris dilatata* (broad buckler-fern).

Selective species: *Fagus sylvatica* (beech), *Acer pseudoplatanus* (sycamore), *Endymion non-scriptus* (bluebell), *Dryopteris filix-mas* (male fern), *Dryopteris dilatata* (broad buckler-fern), *Mnium hornum*

Species groups: A, F (I, E, B, D)

Canopy and understorey species

Constant trees	*Constant saplings*
(oak, sycamore, ash, beech, rowan, birch)	(sycamore, ash, rowan)

Constant shrubs	*Trees (basal area)*
(hazel)	oak (sycamore, beech)

ENVIRONMENT

Geographical distribution	*Solid geology*
(SE, NW, Wa, NE, SW, WS, ME, ES)	Mill grit/Coal mea, Silur/Ordov (E, D, F, C, A, I, B)

Altitude (m)	*Slope (°)*	*Rainfall (cm)*	*Soil pH*	*LOI*
100 *(med)*	14 *(med)*	99 *(med)*	4·3 *(low)*	12·8 *(low)*

GENERAL DESCRIPTION

A type of frequent occurrence in a wide range of site types, with medium heterogeneity and a low species complement, most closely related to types 10, 22 and 24. There is an average ground cover with high litter also. There is a dense canopy with saplings present and an understorey sometimes present also.

This type would probably be termed mixed deciduous woodland and is of variable tree composition. The soils are mainly brown earths, with a tendency towards deposition of iron. The comparable phytosociological associations are probably Ulmo-Fraxinetum E Sjogren ap KL 1973 (Ulmo-Quercetum Tx 1931) and Dryopterido-Fraxinetum Klötzli 1970.

Type 9. *Endymion non-scriptus/Rubus fruticosus* (bluebell/bramble) type
Ground flora species: *Rubus fruticosus* (bramble), *Pteridium aquilinum* (bracken)
Woody species: *Acer pseudoplatanus* (sycamore), *Quercus* spp (oak), *Corylus avellana* (hazel)

–	Absent
•	1 – 4
•	5 – 8
●	9 – 12
●	13 – 16

PLOT TYPE 10

ATHYRIUM FILIX-FEMINA/RUBUS FRUTICOSUS (LADY-FERN/BRAMBLE) TYPE

VEGETATION

Key species

Constant species: *Fraxinus excelsior* (ash), *Rubus fruticosus* (bramble), *Dryopteris dilatata* (broad buckler-fern), *Circaea lutetiana* (enchanter's nightshade), *Dryopteris filix-mas* (male fern), *Quercus* spp (oak), *Corylus avellana* (hazel), *Eurhynchium praelongum*

Plot dominants: *Rubus fruticosus* (bramble)

Selective species: *Athyrium filix-femina* (lady-fern), *Circaea lutetiana* (enchanter's nightshade), *Veronica montana* (wood speedwell), *Dryopteris dilatata* (broad buckler-fern), *Fraxinus excelsior* (ash), *Chrysosplenium oppositifolium* (opposite-leaved golden saxifrage)

Species groups: F, A, B (E, D, I, C, G)

Canopy and understorey species
Constant trees
(ash, oak, sycamore, birch, willow, alder)

Constant saplings
(ash, sycamore)

Constant shrubs
(hazel)

Trees (basal area)
oak, ash

ENVIRONMENT

Geographical distribution
Wa (SW, NW, NE, ME, ES, SE)

Solid geology
Silur/Ordov (G, D, A, B, C, I, F, E)

Altitude (m)	Slope (°)	Rainfall (cm)	Soil pH	LOI
101 (med)	15 (med)	102 (med)	4·8 (low)	13·4 (low)

GENERAL DESCRIPTION

A type of frequent occurrence in a wide range of site types, with medium heterogeneity and an average species complement, most closely related to types 9, 12 and 11. The canopy is usually dense with a few saplings and an understorey is often present. There is an average ground cover with much bare ground below.

This type would probably be termed moist pedunculate oak/ash woodland. The soils are mainly brown earths, although often rather rocky and sometimes shallow. The comparable phytosociological associations are probably Dryopterido-Fraxinetum Klötzli 1970 and Alno-Fraxinetum KL ap Seibert 1969.

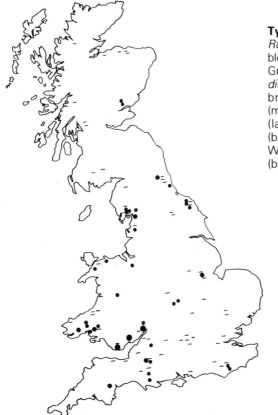

Type 10. *Athyrium filix-femina/ Rubus fruticosus* (lady-fern/bramble) type
Ground flora species: *Brachypodium sylvaticum* (slender false-brome), *Dryopteris filix-mas* (male fern), *Athyrium filix-femina* (lady-fern), *Rubus fruticosus* (bramble)
Woody species: *Betula* spp (birch), *Fraxinus excelsior* (ash)

–	Absent
•	1 – 4
•	5 – 8
●	9 – 12
●	13 – 16

PLOT TYPE 11

POTENTILLA STERILIS/RUBUS FRUTICOSUS (BARREN STRAWBERRY/BRAMBLE) TYPE

VEGETATION

Key species

Constant species: *Rubus fruticosus* (bramble), *Viola riviniana* (common violet), *Quercus* spp (oak), *Fraxinus excelsior* (ash), *Corylus avellana* (hazel), *Betula* spp (birch), *Dryopteris filix-mas* (male fern), *Thuidium tamariscinum*

Plot dominants: *Rubus fruticosus* (bramble)

Selective species: *Potentilla sterilis* (barren strawberry), *Viola riviniana* (common violet), *Fragaria vesca* (wild strawberry), *Prunella vulgaris* (self-heal), *Ajuga reptans* (bugle), *Thuidium tamariscinum*

Species groups: A, E, B, F (D, I, C, J, G)

Canopy and understorey species

Constant trees	Constant saplings
(oak, ash, birch, willow)	(ash, birch, hawthorn, oak)

Constant shrubs	Trees (basal area)
hazel	oak (ash)

ENVIRONMENT

Geographical distribution	Solid geology
SW (NW, SE, Wa, WS, ME, ES)	Silur/Ordov (F, C, D, A, I, E, B, G)

Altitude (m)	Slope (°)	Rainfall (cm)	Soil pH	LOI
94 *(low)*	11 *(med)*	117 *(med)*	5·1 *(med)*	14·1 *(low)*

GENERAL DESCRIPTION

A type of frequent occurrence in a wide range of site types, with medium heterogeneity and an average species complement, most closely related to types 12, 10 and 24. There is usually an average ground cover but with much bare ground. The canopy is quite dense, with an understorey usually present and saplings often present in average density.

This type would probably be included as pedunculate oak woodland or mixed deciduous woodland on lower valley sides. The soils are mainly brown earths. The comparable phytosociological association is probably Dryopterido-Fraxinetum Klötzli 1970.

Type 11. *Potentilla sterilis/ Rubus fruticosus* (barren straw- berry/bramble) type
Ground flora species: *Rubus fruticosus* (bramble), *Bromus ramosus* (hairy brome)
Woody species: *Betula* spp (birch), *Fraxinus excelsior* (ash), *Corylus avellana* (hazel)

–	Absent
•	1 – 4
•	5 – 8
•	9 – 12
●	13 – 16

PLOT TYPE 12

GEUM URBANUM/MERCURIALIS PERENNIS (HERB BENNET/DOG'S MERCURY) TYPE

VEGETATION

Key species

Constant species:
Fraxinus excelsior (ash), *Rubus fruticosus* (bramble), *Viola riviniana* (common violet), *Dryopteris filix-mas* (male fern), *Geum urbanum* (herb bennet), *Oxalis acetosella* (wood-sorrel), *Crataegus monogyna* (common hawthorn), *Deschampsia cespitosa* (tufted hair-grass)

Plct dominants:
Mercurialis perennis (dog's mercury), *Rubus fruticosus* (bramble)

Selective species:
Geum urbanum (herb bennet), *Brachypodium sylvaticum* (slender false-brome), *Fragaria vesca* (wild strawberry), *Sanicula europaea* (sanicle), *Viola riviniana* (common violet), *Deschampsia cespitosa* (tufted hair-grass)

Species groups:
A, F, E, D, B (I, C, G)

Canopy and understorey species

Constant trees	*Constant saplings*
(ash, oak, birch, sycamore, hawthorn, alder)	(ash, hawthorn, sycamore)
Constant shrubs	*Trees (basal area)*
(hazel)	oak (ash)

ENVIRONMENT

Geographical distribution	*Solid geology*
NE (Wa, NW, SW, WS, ME, ES, SE)	Mill grit/Coal mea (E, F, H, I, D, B, C, J)

Altitude (m)	Slope (°)	Rainfall (cm)	Soil pH	LOI
102 *(med)*	19 *(high)*	102 *(med)*	5·3 *(med)*	11·4 *(low)*

GENERAL DESCRIPTION

A type of frequent occurrence in a wide range of site types, with medium heterogeneity and a high species complement, most closely related to types 11, 10 and 13. There is usually low ground cover, with much rock and bare ground. The canopy is quite dense, with an understorey also usually present and with saplings in low density but of frequent occurrence.

This type belongs to the broad range of pedunculate oak-ash woodland growing on quite basiphilous steep valley sides. The soils are mainly brown earths, although often skeletal, very stony and sometimes gleyed. The comparable phytosociological associations are probably Fraxinus-Brachypodium nodum McVean & Ratcliffe 1959 and Alno-Fraxinetum KL ap Seibert 1969.

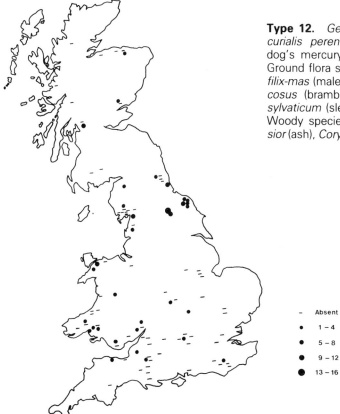

Type 12. *Geum urbanum/Mercurialis perennis* (herb bennet/dog's mercury) type
Ground flora species: *Dryopteris filix-mas* (male fern), *Rubus fruticosus* (bramble), *Brachypodium sylvaticum* (slender false-brome)
Woody species: *Fraxinus excelsior* (ash), *Corylus avellana* (hazel)

–	Absent
•	1 – 4
•	5 – 8
●	9 – 12
●	13 – 16

PLOT TYPE 13

CHRYSOSPLENIUM OPPOSITIFOLIUM/MERCURIALIS PERENNIS (OPPOSITE-LEAVED GOLDEN SAXIFRAGE/DOG'S MERCURY) TYPE

VEGETATION

Key species
Constant species: *Dryopteris filix-mas* (male fern), *Rubus fruticosus* (bramble), *Urtica dioica* (stinging nettle), *Eurhynchium praelongum*

Plot dominants: *Mercurialis perennis* (dog's mercury), *Rubus fruticosus* (bramble), *Urtica dioica* (stinging nettle), *Chrysosplenium oppositifolium* (opposite-leaved golden saxifrage)

Selective species: *Chrysosplenium oppositifolium* (opposite-leaved golden saxifrage), *Silene dioica* (red campion), *Campanula latifolia* (large campanula), *Heracleum sphondylium* (hogweed), *Galium aparine* (goosegrass), *Veronica montana* (wood speedwell)

Species groups: A, F, D (E, B, C, I, G)

Canopy and understorey species
Constant trees
(ash, sycamore, wych elm)

Constant saplings
(ash)

Constant shrubs
(elder)

Trees (basal area)
ash (sycamore, wych elm)

ENVIRONMENT

Geographical distribution
NE (Wa, ES, SW, WS, ME, NW)

Solid geology
Mill grit/Coal mea (H, I, B, D, F, A, J, E)

Altitude (m)	Slope (°)	Rainfall (cm)	Soil pH	LOI
99 (med)	20 (high)	97 (med)	5·5 (med)	11·4 (low)

GENERAL DESCRIPTION

A type of average occurrence in an average range of site types, with medium heterogeneity and an average species complement, most closely related to types 12, 11 and 10. There is usually an average ground cover, with a high proportion of bare ground. The canopy is quite dense, with an infrequent understorey, but with saplings often present in low densities.

This type is mainly ash woodland growing on steep slopes under rich moist conditions. The soils are alluvial brown earths, although there is often some gleying due to frequent waterlogging. The comparable phytosociological association is probably Carici-remotae Fraxinetum (W Koch 1926) Schwickerath 1937.

Type 13. *Chrysosplenium oppositifolium/Mercurialis perennis* (opposite-leaved golden saxifrage/dog's mercury) type
Ground flora species: *Rubus fruticosus* (bramble), *Dryopteris filix-mas* (male fern), *Hedera helix* (ivy), *Mercurialis perennis* (dog's mercury)
Woody species: *Alnus glutinosa* (alder), *Fraxinus excelsior* (ash), *Corylus avellana* (hazel)

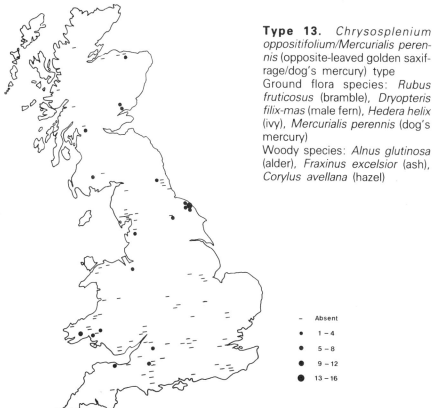

–	Absent
•	1 – 4
•	5 – 8
•	9 – 12
●	13 – 16

PLOT TYPE 14

CHRYSOSPLENIUM OPPOSITIFOLIUM/RUBUS FRUTICOSUS (OPPOSITE-LEAVED GOLDEN SAXIFRAGE/BRAMBLE) TYPE

VEGETATION

Key species

Constant species:
Chrysosplenium oppositifolium (opposite-leaved golden sax-ifrage), *Dryopteris dilatata* (broad buckler-fern), *Silene dioica* (red campion), *Circaea lutetiana* (enchanter's nightshade), *Urtica dioica* (stinging nettle)

Plot dominants:
Rubus fruticosus (bramble), *Chrysosplenium oppositifolium* (opposite-leaved golden saxifrage), *Urtica dioica* (stinging nettle), Filipendula ulmaria (meadow-sweet)

Selective species:
Iris pseudacorus (yellow flag), *Chrysosplenium oppositifolium* (opposite-leaved golden saxifrage), *Alnus glutinosa* (alder), *Galium palustre* (marsh bedstraw), *Solanum dulcamara* (woody nightshade), *Carex paniculata* (panicled sedge)

Species groups:
A, D, F, E, C (G, B, I)

Canopy and understorey species

Constant trees	*Constant saplings*
(alder, ash, willow, birch)	(willow, ash, birch, hawthorn)

Constant shrubs	*Trees (basal area)*
(hazel)	ash, alder

ENVIRONMENT

Geographical distribution	*Solid geology*
Wa (ME, WS, NW, ES, NE, SE)	Red s st, K marl/Lias (D, C, G, H, E)

Altitude (m)	Slope (°)	Rainfall (cm)	Soil pH	LOI
63 *(low)*	6 *(low)*	117 *(med)*	5·5 *(med)*	26·8 *(low)*

GENERAL DESCRIPTION

A type of average occurrence in an average range of site types, with medium heterogeneity and an average species complement, most closely related to types 13, 12 and 16. There is high average ground cover, but with much bare ground and bryophytes. The canopy is usually dense with saplings in high densities and an understorey commonly present.

This type would probably be called mixed ash/alder woodland growing on eutrophic valley floors. The soils are invariably heavily gleyed and often with a surface humus-rich horizon, invariably with surface water. The comparable phytosociological associations are probably Carex laevigatae-Alnetum (Atlantic) Allorge 1922 and Carici-remotae Fraxinetum (W Koch 1926) Schwickerath 1937.

Type 14. *Chrysosplenium oppositifolium/Rubus fruticosus* (opposite-leaved golden saxifrage/bramble) type
Ground flora species: *Juncus effusus* (soft rush), *Athyrium filix-femina* (lady-fern), *Rubus fruticosus* (bramble), *Angelica sylvestris* (wild angelica)
Woody species: *Alnus glutinosa* (alder), *Fraxinus excelsior* (ash)

–	Absent
•	1 – 4
•	5 – 8
•	9 – 12
●	13 – 16

PLOT TYPE 15

GALIUM PALUSTRE/AGROSTIS TENUIS (MARSH BEDSTRAW/COMMON BENT-GRASS) TYPE

VEGETATION

Key species

Constant species: *Holcus lanatus* (Yorkshire fog), *Cirsium palustre* (marsh thistle), *Ranunculus repens* (creeping buttercup), *Rubus fruticosus* (bramble), *Epilobium montanum* (broad-leaved willow-herb), *Juncus effusus* (soft rush), *Prunella vulgaris* (self-heal)

Plot dominants: *Agrostis tenuis* (common bent-grass), *Holcus lanatus* (Yorkshire fog), *Rubus fruticosus* (bramble), *Lolium perenne* (rye-grass), *Agrostis stolonifera* (creeping bent-grass), *Holcus mollis* (creeping soft-grass), *Poa annua* (annual poa)

Selective species: *Hypericum tetrapterum* (square-stemmed St. John's wort), *Senecio aquaticus* (marsh ragwort), *Scrophularia aquatica* (water betony), *Lotus pedunculatus* (large birdsfoot-trefoil), *Centaurea nigra* (lesser knapweed), *Lolium perenne* (rye-grass)

Species groups: A, G, D, E (C, I, F, B, H)

Canopy and understorey species

Constant trees	*Constant saplings*
(oak, alder, ash, willow)	(ash, willow, alder)

Constant shrubs	*Trees (basal area)*
(hazel)	oak (willow)

ENVIRONMENT

Geographical distribution	*Solid geology*
Wa (WS, SW, SE, ME, NE)	Devonian, Silur/Ordov (G, I, C)

Altitude (m)	*Slope (°)*	*Rainfall (cm)*	*Soil pH*	*LOI*
80 *(low)*	7 *(low)*	119 *(med)*	5·4 *(med)*	12·2 *(low)*

GENERAL DESCRIPTION

A type of infrequent occurrence but in an average range of site types, with medium heterogeneity and a high species complement, most closely related to types 16, 31 and 14, with a dense ground cover often grazed by cattle, and a high cover of litter. The canopy is usually open with few saplings or shrubs, without an understorey.

This type would, in many cases, be called scrub or occasionally grassland often under moist, heavy conditions. The soils are mainly gleyed brown earths in recent alluvium, but some brown earths are also present. The comparable phytosociological association is probably Alno-Fraxinetum KL ap Seibert 1969, but transitions are also present with grassland associations.

Type 15. *Galium palustre/ Agrostis tenuis* (marsh bedstraw/ common bent-grass) type
Ground flora species: *Holcus lanatus* (Yorkshire fog), *Rubus fruticosus* (bramble)
Woody species: *Corylus avellana* (hazel)

–	Absent
•	1 – 4
•	5 – 8
•	9 – 12
●	13 – 16

PLOT TYPE 16

CIRSIUM PALUSTRE/AGROSTIS TENUIS (MARSH THISTLE/COMMON BENT-GRASS) TYPE

VEGETATION

Key species
Constant species: *Cirsium palustre* (marsh thistle), *Fraxinus excelsior* (ash), *Viola riviniana* (common violet), *Athyrium filix-femina* (lady-fern), *Circaea lutetiana* (enchanter's nightshade), *Deschampsia cespitosa* (tufted hair-grass), *Agrostis tenuis* (common bent-grass), *Holcus lanatus* (Yorkshire fog), *Lysimachia nemorum* (yellow pimpernel)

Plot dominants: *Agrostis tenuis* (common bent-grass)

Selective species: *Galium mollugo* (hedge bedstraw), *Cirsium palustre* (marsh thistle), *Cerastium holosteoides* (common mouse-ear chickweed), *Carex binervis* (ribbed sedge), *Cynosurus cristatus* (crested dog's-tail), *Dactylis glomerata* (cock's-foot)

Species groups: A, E, I, G, D (C, F, H, J, B)

Canopy and understorey species

| *Constant trees* | *Constant saplings* |
| ash (alder) | (ash) |

| *Constant shrubs* | *Trees (basal area)* |
| (hazel) | ash (alder) |

ENVIRONMENT

| *Geographical distribution* | *Solid geology* |
| Wa (NW, NE, WS, SW, ES) | Silur/Ordov (G, I, J, D, F) |

Altitude *(m)*	Slope *(°)*	Rainfall *(cm)*	Soil pH	LOI
140 *(med)*	23 *(high)*	150 *(high)*	5·1 *(med)*	13·8 *(low)*

GENERAL DESCRIPTION

A type of infrequent occurrence in an average range of site types, with high heterogeneity and a high species complement, most closely related to types 30, 29 and 11. There is average ground cover, often grazed by sheep, with a high ground cover of rock. The canopy is open, with few saplings or shrubs.

This type would probably be called riverine, upland ash woodland'. The soils are usually gleys flushed by lateral water movement and are often shallow. The comparable phytosociological association is probably Carici-remotae (W Koch 1926) Schwickerath 1937, but transitions are also present with grassland associations.

Type 16. *Cirsium palustre/ Agrostis tenuis* (marsh thistle/ common bent-grass) type
Ground flora species: *Agrostis tenuis* (common bent-grass), *Anthoxanthum odoratum* (sweet vernal-grass), *Oxalis acetosella* (wood-sorrel)
Woody species: *Fraxinus excelsior* (ash), *Corylus avellana* (hazel)

Absent

• 1 – 4

• 5 – 8

● 9 – 12

● 13 – 16

71

PLOT TYPE 17

PTERIDIUM AQUILINUM/RUBUS FRUTICOSUS (BRACKEN/BRAMBLE) TYPE

VEGETATION

Key species
Constant species:　　　*Quercus* spp (oak), *Rubus fruticosus* (bramble)

Plot dominants:　　　　*Rubus fruticosus* (bramble), *Pteridium aquilinum* (bracken), *Hedera helix* (ivy)

Selective species:　　　*Ilex aquifolium* (holly), *Fagus sylvatica* (beech), *Castanea sativa* (Spanish chestnut), *Carpinus betulus* (hornbeam), *Isopterygium elegans, Dicranella heteromalla*

Species groups:　　　　(F, B, I)

Canopy and understorey species

Constant trees (oak, beech, birch, holly, rowan)	*Constant saplings* (holly, beech)
Constant shrubs —	*Trees (basal area)* oak (beech)

ENVIRONMENT

Geographical distribution SE (ME, NE, Wa, SW, NW)	*Solid geology* Wealden (G, E, A)

Altitude (m)	*Slope (°)*	*Rainfall (cm)*	*Soil pH*	*LOI*
108 *(med)*	9 *(low)*	81 *(low)*	3·9 *(low)*	18·8 *(low)*

GENERAL DESCRIPTION

A type of frequent occurrence but in a restricted range of site types, with very low heterogeneity and a low species complement, most closely related to types 23, 21 and 20. There is very low ground cover, but very high cover of leaf litter. The canopy is mainly dense with few saplings or shrubs.

This type would probably be termed dry, acid sessile oak woodland. The soils are mainly acid brown earths. The comparable phytosociological associations are probably Blechno-Quercetum Br-Bl et Tx 1952 and Fago-Quercetum petraeae Tx 1955.

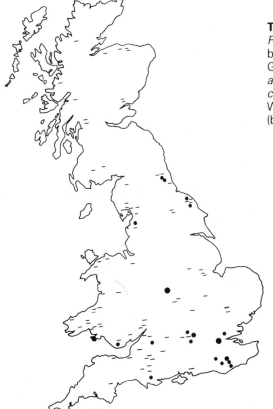

Type 17. *Pteridium aquilinum/ Rubus fruticosus* (bracken/bramble) type
Ground flora species: *Pteridium aquilinum* (bracken), *Rubus fruticosus* (bramble)
Woody species: *Fagus sylvatica* (beech), *Quercus* spp (oak)

–	Absent
•	1 – 4
•	5 – 8
●	9 – 12
●	13 – 16

PLOT TYPE 18

DESCHAMPSIA FLEXUOSA/PTERIDIUM AQUILINUM (WAVY HAIR-GRASS/BRACK-EN) TYPE

VEGETATION

Key species

Constant species: *Quercus* spp (oak), *Deschampsia flexuosa* (wavy hair-grass), *Pteridium aquilinum* (bracken), *Betula* spp (birch), *Lonicera periclymenum* (honeysuckle)

Plot dominants: *Pteridium aquilinum* (bracken), *Deschampsia flexuosa* (wavy hair-grass), *Vaccinium myrtillus* (bilberry), *Rubus fruticosus* (bramble)

Selective species: *Deschampsia flexuosa* (wavy hair-grass), *Vaccinium myrtillus* (bilberry), *Lonicera periclymenum* (honeysuckle), *Leucobryum glaucum, Dicranella heteromalla, Dicranum scoparium*

Species groups: I, F (B, J)

Canopy and understorey species

Constant trees	*Constant saplings*
oak (birch, rowan)	(birch, oak, rowan)
Constant shrubs	*Trees (basal area)*
—	oak (birch)

ENVIRONMENT

Geographical distribution	*Solid geology*
SE (NW, Wa, NE, SW, ES)	Silur/Ordov, Wealden (G, B, E, I)

Altitude (m)	Slope (°)	Rainfall (cm)	Soil pH	LOI
122 (med)	12 (med)	109 (med)	3·9 (low)	24·7 (low)

GENERAL DESCRIPTION

A type of average occurrence in an average range of site types, with low heterogeneity and a low species complement, most closely related to types 25, 26 and 23. There is average ground cover and high cover of litter. The canopy is of medium density, with few saplings or shrubs.

This type would probably be termed dry acid sessile oak or birch woodland. The soils are mainly brown podzolic or are podzolic in character. The comparable phytosociological associations are probably Galio saxatilis-Quercetum Birse et Robertson 1976 and Blechno-Quercetum Br-Bl et Tx 1952 or Betulo-Quercetum Tx 1937.

Type 18. *Deschampsia flex-uosa/Pteridium aquilinum* (wavy hair-grass/bracken) type
Ground flora species: *Pteridium aquilinum* (bracken), *Teucrium scorodonia* (wood sage)
Woody species: *Quercus* spp (oak), *Betula* spp (birch)

- Absent
• 1 – 4
• 5 – 8
● 9 – 12
● 13 – 16

75

CHAMAENERION ANGUSTIFOLIUM/PTERIDIUM AQUILINUM (ROSEBAY WIL-
LOW-HERB/BRACKEN) TYPE

VEGETATION

Key species
Constant species: *Rubus fruticosus* (bramble), *Chamaenerion angustifolium*
                        (rosebay willow-herb)

Plot dominants: *Pteridium aquilinum* (bracken), *Rubus fruticosus* (bramble)

Selective species: *Chamaenerion angustifolium* (rosebay willow-herb), *Rho-*
                        dodendron ponticum (rhododendron), *Ulex europaeus*
                        (gorse), *Holcus lanatus* (Yorkshire fog), *Rubus fruticosus*
                        (bramble), *Dicranella heteromalla*

Species groups: (F, I, B, G, E, J, A)

Canopy and understorey species
Constant trees   *Constant saplings*
(birch, oak, sycamore)   (birch)

Constant shrubs   *Trees (basal area)*
Rhododendron   (oak, birch)

ENVIRONMENT

Geographical distribution   *Solid geology*
SE, SW (NW, ME, NE)   Wealden, Devonian (E, G, B, A, F, H)

Altitude (m)	Slope (°)	Rainfall (cm)	Soil pH	LOI
117 *(med)*	15 *(med)*	94 *(med)*	4·3 *(low)*	16·6 *(low)*

GENERAL DESCRIPTION

A type of infrequent occurrence in an average range of site types, with medium
heterogeneity and a low species complement, most closely related to types 20, 21
and 22. There is very high ground cover, with a high litter cover also. The canopy is
usually open, with a few shrubs or saplings.

   This type is composed mainly of scrub on lowland heaths with much bracken, and
would probably therefore be called heathy birch scrub. The soils are mainly acid
brown earths and are usually very dry and shallow. The comparable phytosociological
association is probably Betulo-Quercetum Tx 1937, but other heath associations may
also be involved.

Type 19. *Chamaenerion angustifolium/Pteridium aquilinum* (rosebay willow-herb/bracken) type
Ground flora species: *Pteridium aquilinum* (bracken), *Vaccinium myrtillus* (bilberry)
Woody species: *Quercus* spp (oak), *Betula* spp (birch)

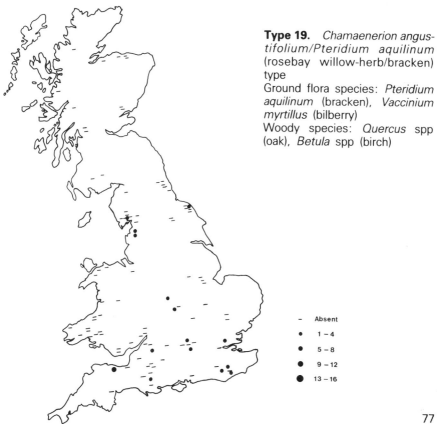

–	Absent
•	1 – 4
•	5 – 8
●	9 – 12
⬤	13 – 16

PLOT TYPE 20

CHAMAENERION ANGUSTIFOLIUM/RUBUS FRUTICOSUS (ROSEBAY WILLOW-HERB/BRAMBLE) TYPE

VEGETATION

Key species

Constant species: *Rubus fruticosus* (bramble), *Holcus lanatus* (Yorkshire fog), *Quercus* spp *(oak)*

Plot dominants: *Rubus fruticosus* (bramble), *Pteridium aquilinum* (bracken), *Deschampsia flexuosa* (wavy hair-grass)

Selective species: *Chamaenerion angustifolium* (rosebay willow-herb), *Carpinus betulus* (hornbeam), *Holcus lanatus* (Yorkshire fog), *Fagus sylvatica* (beech), *Juncus effusus* (soft rush), *Dicranella heteromalla*

Species groups: (I, E, B, F, G, A, D, J)

Canopy and understorey species

Constant trees
(beech, oak, birch)

Constant saplings
(birch, oak)

Constant shrubs
—

Trees (basal area)
beech, oak

ENVIRONMENT

Geographical distribution
SE (NW, SW, Wa, ME, NE)

Solid geology
Wealden, Oolite/Chalk (G, H, D, B, I)

Altitude (m)	Slope (°)	Rainfall (cm)	Soil pH	LOI
123 *(med)*	6 *(low)*	81 *(low)*	4·0 *(low)*	10·4 *(low)*

GENERAL DESCRIPTION

A type of average occurrence in an average range of site types, with medium heterogeneity and a low species complement, most closely related to types 21, 19 and 23. There is low ground cover, with a high leaf litter cover. The canopy is of average density, with few saplings or shrubs.

This type would probably be termed mixed sessile oak/beech woodland on acid soils. The soils are mainly acid brown earths and are often rather heavy. The comparable phytosociological associations are probably Deschampsio-Fagetum Passarge 1956 and Fago-Quercetum Tx 1955.

Type 20. *Chamaenerion angustifolium/Rubus fruticosus* (rosebay willow-herb/bramble) type
Ground flora species: *Pteridium aquilinum* (bramble), *Agrostis tenuis* (common bent-grass)
Woody species: *Fagus sylvatica* (beech), *Castanea sativa* (sweet chestnut)

–	Absent
•	1 – 4
•	5 – 8
●	9 – 12
●	13 – 16

PLOT TYPE 21

OXALIS ACETOSELLA/PTERIDIUM AQUILINUM (WOOD-SORREL/BRACKEN) TYPE

VEGETATION

Key species

Constant species: *Oxalis acetosella* (wood-sorrel), *Dryopteris dilatata* (broad buckler-fern), *Rubus fruticosus* (bramble), *Mnium hornum*

Plot dominants: *Pteridium aquilinum* (bracken), *Rubus fruticosus* (bramble), *Holcus mollis* (creeping soft-grass), *Dryopteris dilatata* (broad buckler-fern)

Selective species: *Milium effusum* (wood millet), *Oxalis acetosella* (wood-sorrel), *Digitalis purpurea* (foxglove), *Larix europaea* (European larch), *Acer pseudoplatanus* (sycamore), *Mnium hornum*

Species groups: F, I (E, B, A, G, J)

Canopy and understorey species

Constant trees	*Constant saplings*
(oak, birch, sycamore, larch)	—

Constant shrubs	*Trees (basal area)*
(hazel)	oak (sycamore, larch)

ENVIRONMENT

Geographical distribution	*Solid geology*
NW, NE (Wa, SE, ES, ME, SW)	Mill grit/Coal mea, Silur/Ordov (I, F, D, A, B, C, E)

Altitude (m)	*Slope (°)*	*Rainfall (cm)*	*Soil pH*	*LOI*
101 *(med)*	14 *(med)*	107 *(med)*	4·0 *(low)*	18·6 *(low)*

GENERAL DESCRIPTION

A type of average occurrence in a wide range of site types, with medium heterogeneity and a low species complement, most closely related to types 20, 19 and 23. There is a low ground cover, with litter predominating. The canopy is usually dense, with few saplings or shrubs.

This type would mainly be called acid sessile oak woodland on valley sides. The soils are usually acid brown earths. The comparable phytosociological associations are probably Blechno-Quercetum Br-Bl et Tx 1952 or Fago-Quercetum Tx 1955.

Type 21. *Oxalis acetosella/ Pteridium aquilinum* (wood-sorrel/bracken) type
Ground flora species: *Pteridium aquilinum* (bracken)
Woody species: *Quercus* spp (oak), *Betula* spp (birch)

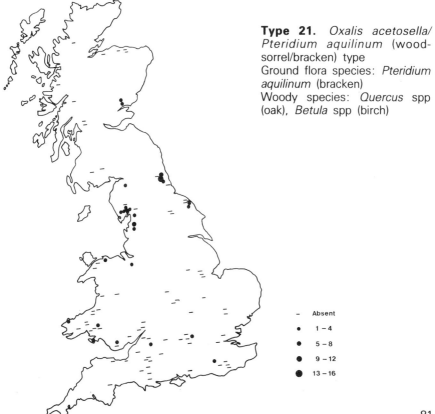

–	Absent
•	1 – 4
•	5 – 8
●	9 – 12
●	13 – 16

PLOT TYPE 22

BLECHNUM SPICANT/RUBUS FRUTICOSUS (HARD-FERN/BRAMBLE) TYPE

VEGETATION

Key species

Constant species: *Dryopteris dilatata* (broad buckler-fern), *Dryopteris filix-mas* (male fern), *Quercus* spp (oak), *Oxalis acetosella* (wood-sorrel), *Rubus fruticosus* (bramble), *Corylus avellana* (hazel), *Mnium hornum*

Plot dominants: *Rubus fruticosus* (bramble), *Luzula sylvatica* (greater wood-rush), *Holcus mollis* (creeping soft-grass)

Selective species: *Blechnum spicant* (hard-fern), *Luzula sylvatica* (greater wood-rush), *Sorbus aucuparia* (rowan), *Fagus sylvatica* (beech), *Athyrium filix-femina* (lady-fern), *Mnium hornum*

Species groups: F, I (A, B, E, J, D, G)

Canopy and understorey species

Constant trees	*Constant saplings*
oak, (birch, rowan, beech, ash, sycamore)	(rowan, birch, ash)

Constant shrubs	*Trees (basal area)*
(hazel)	oak

ENVIRONMENT

Geographical distribution	*Solid geology*
Wa (NE, ES, NW, SW, WS, ME)	Silur/Ordov, Mill grit/Coal mea, Red s st (D, C, A, B, E, J)

Altitude (m)	Slope (°)	Rainfall (cm)	Soil pH	LOI
112 *(med)*	22 *(high)*	109 *(med)*	4·3 *(low)*	17·1 *(low)*

GENERAL DESCRIPTION

A type of average occurrence in a wide range of site types, with medium heterogeneity and an average species complement, most closely related to types 19, 21 and 23. There is an average ground cover, with high litter cover. The canopy is usually of average density, with few saplings and a variable quantity of shrubs.

This type would probably be called mixed deciduous woodland on steep valley sides. The soils are mainly acid brown earths and are often very rocky. The comparable phytosociological associations are probably Blechno-Quercetum Br-Bl et Tx (1950) 1952 and Fago-Quercetum Tx 1955.

Type 22. *Blechnum spicant/ Rubus fruticosus* (hard-fern/ bramble) type
Ground flora species: *Luzula sylvatica* (greater woodrush), *Rubus fruticosus* (bramble)
Woody species: *Quercus* spp (oak), *Corylus avellana* (hazel)

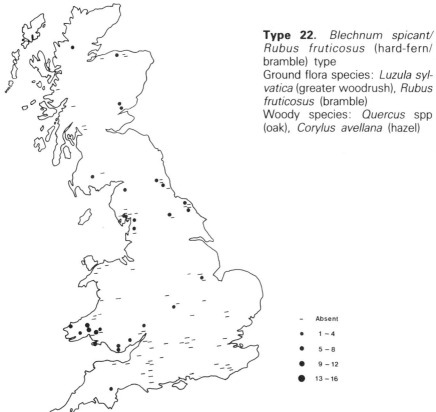

–	Absent
•	1 – 4
●	5 – 8
●	9 – 12
●	13 – 16

PLOT TYPE 23

HOLCUS MOLLIS/PTERIDIUM AQUILINUM (CREEPING SOFT-GRASS/BRACKEN)
TYPE

VEGETATION

Key species
Constant species: *Rubus fruticosus* (bramble), *Quercus* spp (oak), *Pteridium aquilinum* (bracken), *Oxalis acetosella* (wood-sorrel), *Holcus mollis* (creeping soft-grass)

Plot dominants: *Pteridium aquilinum* (bracken), *Rubus fruticosus* (bramble), *Holcus mollis* (creeping soft-grass), *Dryopteris dilatata* (broad buckler-fern)

Selective species: *Holcus mollis* (creeping soft-grass), *Pteridium aquilinum* (bracken), *Endymion non-scriptus* (bluebell), *Oxalis acetosella* (wood-sorrel), *Stellaria holostea* (greater stitchwort), *Quercus* spp (oak)

Species groups: F (I, B, E)

Canopy and understorey species

Constant trees	*Constant saplings*
oak (birch)	—

Constant shrubs	*Trees (basal area)*
(hazel)	oak

ENVIRONMENT

Geographical distribution	*Solid geology*
SW (SE, NW, NE, Wa, ME)	Devonian, Wealden (H, I, G, B, E, A)

Altitude (m)	*Slope (°)*	*Rainfall (cm)*	*Soil pH*	*LOI*
124 *(med)*	11 *(med)*	107 *(med)*	4·0 *(low)*	14·6 *(low)*

GENERAL DESCRIPTION

A type of average occurrence in an average range of site types, with low heterogeneity and a low species complement, most closely related to types 21, 17 and 22. There is a high average ground cover, with much leaf litter. The canopy is invariably dense, with few saplings and some shrubs.

This type would probably be termed acidic oak/birch woodland on rather heavy soils. The soils are mainly acid brown earths, often with some evidence of gleying. The comparable phytosociological associations are probably Blechno-Quercetum Br-Bl et Tx 1952 and Galio saxatilis-Quercetum Birse et Robertson 1976 or Betulo-Quercetum Tx 1937.

Type 23. *Holcus mollis/Pteridium aquilinum* (creeping soft-grass/bracken) type
Ground flora species: *Holcus mollis* (creeping soft-grass), *Rubus fruticosus* (bramble), *Teucrium scorodonia* (wood sage)
Woody species: *Quercus* spp (oak), *Corylus avellana* (hazel)

–   Absent
●   1 – 4
●   5 – 8
●   9 – 12
●   13 – 16

PLOT TYPE 24

LUZULA PILOSA/PTERIDIUM AQUILINUM (HAIRY WOODRUSH/BRACKEN) TYPE

VEGETATION

Key species

Constant species: *Rubus fruticosus* (bramble), *Lonicera periclymenum* (honeysuckle), *Quercus* spp (oak), *Pteridium aquilinum* (bracken), *Corylus avellana* (hazel)

Plot dominants: *Pteridium aquilinum* (bracken), *Rubus fruticosus* (bramble), *Holcus mollis* (creeping soft-grass)

Selective species: *Luzula pilosa* (hairy woodrush), *Stellaria holostea* (greater stitchwort), *Lonicera periclymenum* (honeysuckle), *Festuca gigantea* (tall brome), *Holcus mollis* (creeping soft-grass), *Pteridium aquilinum* (bracken)

Species groups: F (B, I, E, A, D)

Canopy and understorey species

Constant trees	*Constant saplings*
oak (birch, hawthorn, beech)	(oak, hawthorn)

Constant shrubs	*Trees (basal area)*
hazel	oak, birch

ENVIRONMENT

Geographical distribution	*Solid geology*
ME, SW (SE, NE)	K marl/Lias (C, D, A, G)

Altitude (m)	Slope (°)	Rainfall (cm)	Soil pH	LOI
83 *(low)*	4 *(low)*	79 *(low)*	4·3 *(low)*	10·4 *(low)*

GENERAL DESCRIPTION

A type of infrequent occurrence in a restricted range of site types, with medium heterogeneity and a low species complement, most closely related to types 22, 11 and 9. There is a high average ground cover, with much leaf litter. There is usually an average canopy beneath which saplings are present in low density, with an understorey often present.

This type would probably be termed acidic oak/birch woodland on heavy soils. The soils are mainly gleys or gleyed brown earths. The comparable phytosociological associations are probably Betulo-Quercetum Tx 1937 or Fago-Quercetum Tx 1955.

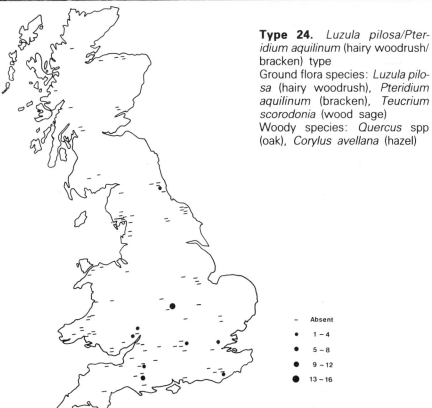

Type 24. *Luzula pilosa/Pteridium aquilinum* (hairy woodrush/bracken) type

Ground flora species: *Luzula pilosa* (hairy woodrush), *Pteridium aquilinum* (bracken), *Teucrium scorodonia* (wood sage)

Woody species: *Quercus* spp (oak), *Corylus avellana* (hazel)

–	Absent
●	1 – 4
●	5 – 8
●	9 – 12
●	13 – 16

PLOT TYPE 25

GALIUM SAXATILE/DESCHAMPSIA FLEXUOSA (HEATH BEDSTRAW/WAVY HAIR-GRASS) TYPE

VEGETATION

Key species
Constant species: *Quercus* spp (oak), *Deschampsia flexuosa* (wavy hair-grass), *Galium saxatile* (heath bedstraw), *Pteridium aquilinum* (bracken), *Oxalis acetosella* (wood-sorrel), *Anthoxanthum odoratum* (sweet vernal-grass), *Betula* spp (birch), *Sorbus aucuparia* (rowan)

Plot dominants: *Deschampsia flexuosa* (wavy hair-grass), *Pteridium aquilinum* (bracken), *Agrostis tenuis* (common bent-grass), *Anthoxanthum odoratum* (sweet vernal-grass)

Selective species: *Galium saxatile* (heath bedstraw), *Deschampsia flexuosa* (wavy hair-grass), *Agrostis canina* (brown bent-grass), *Anthoxanthum odoratum* (sweet vernal-grass), *Vaccinium myrtillus* (bilberry), *Dicranum scoparium*

Species groups: I, F, J (B, E, G)

Canopy and understorey species

Constant trees	*Constant saplings*
oak (birch)	(birch, oak)

Constant shrubs	*Trees (basal area)*
—	oak

ENVIRONMENT

Geographical distribution	*Solid geology*
Wa, NW (WS, SW, NE, SE)	Silur/Ordov (G, D, J, C, I)

Altitude (m)	*Slope (°)*	*Rainfall (cm)*	*Soil pH*	*LOI*
150 *(high)*	23 *(high)*	142 *(high)*	4·1 *(low)*	27·1 *(low)*

GENERAL DESCRIPTION

A type of frequent occurrence in an average range of site types, with medium heterogeneity and a low species complement, most closely related to types 26, 18 and 27. There is average ground cover, but with high rock and bryophyte cover present. The canopy is usually of average density, with few saplings and an understorey rarely present.

This type would be called western acid sessile oakwood. The soils are mainly brown podzolic in character. The comparable phytosociological associations are probably Galio saxatilis-Quercetum Birse et Robertson 1976 and Blechno-Quercetum Br-Bl et Tx 1952 or Betulo-Quercetum Tx 1937.

Type 25. *Galium saxatile/Deschampsia flexuosa* (heath bedstraw/wavy hair-grass) type
Ground flora species: *Pteridium aquilinum* (bracken), *Anthoxanthum odoratum* (sweet vernalgrass), *Deschampsia flexuosa* (wavy hair-grass)
Woody species: *Quercus* spp (oak), *Betula* spp (birch)

–	Absent
•	1 – 4
•	5 – 8
●	9 – 12
●	13 – 16

PLOT TYPE 26

POTENTILLA ERECTA/HOLCUS MOLLIS (COMMON TORMENTIL/CREEPING SOFT-GRASS) TYPE

VEGETATION

Key species

Constant species: *Oxalis acetosella* (wood-sorrel), *Deschampsia flexuosa* (wavy hair-grass), *Pteridium aquilinum* (bracken), *Galium saxatile* (heath bedstraw), *Sorbus aucuparia* (rowan), *Viola riviniana* (common violet), *Anthoxanthum odoratum* (sweet vernal-grass), *Dryopteris dilatata* (broad buckler-fern), *Potentilla erecta* (common tormentil)

Plot dominants: *Holcus mollis* (creeping soft-grass), *Pteridium aquilinum* (bracken), *Agrostis tenuis* (common bent-grass), *Deschampsia flexuosa* (wavy hair-grass), *Luzula sylvatica* (greater woodrush)

Selective species: *Potentilla erecta* (common tormentil), *Galium saxatile* (heath bedstraw), *Anthoxanthum odoratum* (sweet vernal-grass), *Deschampsia flexuosa* (wavy hair-grass), *Vaccinium myrtillus* (bilberry), *Dicranum scoparium*

Species groups: I, J, F, E (D, A, G, B, C, K)

Canopy and understorey species

Constant trees
(oak, birch, rowan)

Constant saplings
(rowan, birch)

Constant shrubs
(hazel)

Trees (basal area)
oak (birch)

ENVIRONMENT

Geographical distribution
Wa (ES, WS, NW, NE)

Solid geology
Silur/Ordov, Red s st, Ign/Metam (G, C, F)

Altitude (m)	Slope (°)	Rainfall (cm)	Soil pH	LOI
128 *(med)*	23 *(high)*	150 *(high)*	4·5 *(low)*	23·1 *(low)*

GENERAL DESCRIPTION

A type of average occurrence in a wide range of site types, with medium heterogeneity and an average species complement, most closely related to types 25, 29 and 30. There is average ground cover, with a high ground cover of rock and bryophytes present. The canopy is usually of average density, with saplings and an understorey often present.

This type would be usually termed western acid sessile oak woodland but has some enrichment. The soils vary in character from acid brown earths to brown podzolic types. The comparable phytosociological associations are probably Galio saxatilis-Quercetum Birse et Robertson 1976 or Blechno-Quercetum Br-Bl et Tx (1950) 1952.

Type 26. *Potentilla erecta/Holcus mollis* (common tormentil/ creeping soft-grass) type
Ground flora species: *Dryopteris filix-mas* (male fern), *Agrostis tenuis* (common bent-grass), *Anthoxanthum odoratum* (sweet vernal-grass)
Woody species: *Corylus avellana* (hazel)

–	Absent
•	1 – 4
•	5 – 8
•	9 – 12
●	13 – 16

PLOT TYPE 27

CALLUNA VULGARIS/PTERIDIUM AQUILINUM (LING/BRACKEN) TYPE

VEGETATION

Key species

Constant species: *Deschampsia flexuosa* (wavy hair-grass), *Galium saxatile* (heath bedstraw), *Pteridium aquilinum* (bracken), *Betula* spp (birch), *Calluna vulgaris* (ling), *Anthoxanthum odoratum* (sweet vernal-grass), *Potentilla erecta* (common tormentil), *Sorbus aucuparia* (rowan), *Vaccinium myrtillus* (bilberry)

Plot dominants: *Pteridium aquilinum* (bracken), *Deschampsia flexuosa* (wavy hair-grass), *Calluna vulgaris* (ling), *Agrostis canina* (brown bent-grass), *Agrostis tenuis* (common bent-grass), *Vaccinium myrtillus* (bilberry)

Selective species: *Calluna vulgaris* (ling), *Erica cinerea* (bell-heather), *Luzula multiflora* (many-headed woodrush), *Vaccinium myrtillus* (bilberry), *Pleurozium schreberi, Hylocomium splendens*

Species groups: J, I (K, F, E, G, D)

Canopy and understorey species

Constant trees	Constant saplings
(birch, oak)	(birch)

Constant shrubs	Trees (basal area)
—	oak (birch)

ENVIRONMENT

Geographical distribution
ES, WS (Wa, NW, SE, ME)

Solid geology
Ign/Metam, Silur/Ordov (I, C)

Altitude (m)	Slope (°)	Rainfall (cm)	Soil pH	LOI
179 (high)	20 (high)	127 (high)	4·4 (low)	26·2 (low)

GENERAL DESCRIPTION

A type of average occurrence in a wide range of site types, with medium heterogeneity and an average species complement, most closely related to types 25, 26 and 18. There is usually a high average ground cover, often grazed by sheep, with a high cover of rock and bryophytes. The canopy is rather open and there is rarely an understorey or saplings present.

This type covers a range of traditional descriptions but is mainly birch or oak woodland in poor upland freely-drained conditions. The soils are mainly brown podzolic or podzols. The comparable phytosociological associations are probably Galio saxatilis-Quercetum Birse et Robertson 1976 or Betulo-Quercetum Tx 1937.

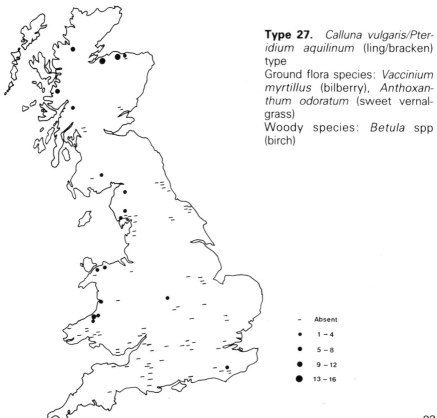

Type 27. *Calluna vulgaris/Pter-idium aquilinum* (ling/bracken) type
Ground flora species: *Vaccinium myrtillus* (bilberry), *Anthoxan-thum odoratum* (sweet vernal-grass)
Woody species: *Betula* spp (birch)

–	Absent
•	1 – 4
•	5 – 8
•	9 – 12
●	13 – 16

PLOT TYPE 28

NARTHECIUM OSSIFRAGUM/MOLINIA CAERULEA (BOG ASPHODEL/PURPLE MOOR-GRASS)

VEGETATION

Key species

Constant species: *Molinia caerulea* (purple moor-grass), *Potentilla erecta* (common tormentil), *Agrostis canina* (brown bent-grass), *Betula* spp (birch), *Carex echinata* (star sedge), *Calluna vulgaris* (ling), *Sorbus aucuparia* (rowan), *Sphagnum* spp

Plot dominants: *Molinia caerulea* (purple moor-grass), *Pteridium aquilinum* (bracken), *Calluna vulgaris* (ling), *Vaccinium myrtillus* (bilberry)

Selective species: *Narthecium ossifragum* (bog asphodel), *Carex echinata* (star sedge), *Drosera rotundifolia* (sundew), *Trichophorum cespitosum* (deer-grass), *Eriophorum angustifolium* (common cotton-grass), *Erica tetralix* (cross-leaved heath)

Species groups: J, I (K, F, G)

Canopy and understorey species

| *Constant trees* | *Constant saplings* |
| (birch, Scots pine) | (birch) |

| *Constant shrubs* | *Trees (basal area)* |
| — | (birch, Scots pine) |

ENVIRONMENT

| *Geographical distribution* | *Solid geology* |
| WS (ES, ME, NW, Wa) | Ign/Metam (C, H) |

Altitude *(m)*	Slope *(°)*	Rainfall *(cm)*	Soil pH	LOI
191 *(high)*	11 *(med)*	152 *(high)*	4·3 *(low)*	63·1 *(high)*

GENERAL DESCRIPTION

A type of infrequent occurrence in a restricted range of sites, with medium heterogeneity and an average species complement, most closely related to types 27, 25 and 26. There is an average ground cover and a high cover of bryophytes. The canopy is usually very open, beneath which neither saplings nor shrubs are commonly present.

This type contains most native pinewood and upland birch woods. The soils are mainly peaty podzols, peaty gleys or peats. As such, they overlap a series of phytosociological associations, including Vaccinio uliginosi-Pinetum Kleist 1929, or Vaccinio-Pinetum Caj 1921, or Barbilophozio-Pinetum Br-Bl et Siss 1939 em Kl 1967, or Betulion pubescentis Lohm et Tx ap Tx 1955 em Scamoni et Passarge 1955.

Type 28. *Narthecium ossifra-gum/Molinia caerulea* (bog asphodel/purple moor-grass) type

Ground flora species: *Narthecium ossifragum* (bog asphodel), *Thelypteris oreopteris* (mountain fern), *Juncus articulatus* (jointed rush), *Pteridium aquilinum* (bracken)

Woody species: *Betula* spp (birch), *Pinus sylvestris* (Scots pine)

- Absent
• 1 – 4
• 5 – 8
● 9 – 12
● 13 – 16

PLOT TYPE 29

ANTHOXANTHUM ODORATUM/AGROSTIS TENUIS (SWEET VERNAL-GRASS/
COMMON BENT-GRASS) TYPE

VEGETATION

Key species

Constant species: *Anthoxanthum odoratum* (sweet vernal-grass), *Oxalis acetosella* (wood-sorrel), *Agrostis tenuis* (common bent-grass), *Rubus fruticosus* (bramble)

Plot dominants: *Agrostis tenuis* (common bent-grass), *Pteridium aquilinum* (bracken), *Holcus mollis* (creeping soft-grass), *Holcus lanatus* (Yorkshire fog)

Selective species: *Anthoxanthum odoratum* (sweet vernal-grass), *Rumex acetosa* (sorrel), *Potentilla erecta* (common tormentil), *Galium saxatile* (heath bedstraw), *Agrostis tenuis* (common bent-grass), *Rhytidiadelphus squarrosus*

Species groups: I, E, F (D, G, A, J, C, B)

Canopy and understorey species

Constant trees	*Constant saplings*
(oak, birch, alder)	—

Constant shrubs	*Trees (basal area)*
(hazel)	oak (birch)

ENVIRONMENT

Geographical distribution	*Solid geology*
Wa (NW, WS, NE, SW, ME)	Silur/Ordov (G, I, J, C, D, F)

Altitude (m)	*Slope (°)*	*Rainfall (cm)*	*Soil pH*	*LOI*
122 *(med)*	19 *(high)*	140 *(high)*	4·5 *(low)*	20·9 *(low)*

GENERAL DESCRIPTION

A type of average occurrence in an average range of site types, with medium heterogeneity and an average species complement, most closely related to types 30, 16 and 26. There is average ground cover, but with a high cover of rocks and bryophytes. The canopy is usually of average density, with saplings and an understorey often present.

This type would probably be called upland oak or birch woodland. The soils are variable, but are mainly acid brown earths and brown podzolic in character. The comparable phytosociological associations are probably Galio saxatilis-Quercetum Birse et Robertson 1976 and Betulion pubescentis Lohm et Tx ap Tx 1955 em Scamoni et Passarge 1959 or Blechno-Quercetum Br-Bl et Tx 1952.

Type 29. *Anthoxanthum odor-atum/Agrostis tenuis* (sweet vernal-grass/common bent-grass) type
Ground flora species: *Rubus fruticosus* (bramble), *Dryopteris filix-mas* (male fern), *Agrostis tenuis* (common bent-grass)
Woody species: *Fraxinus excelsior* (ash)

-   Absent
- • 1 – 4
- • 5 – 8
- • 9 – 12
- ● 13 – 16

PLOT TYPE 30

SUCCISA PRATENSIS/HOLCUS MOLLIS (DEVIL'S-BIT SCABIOUS/CREEPING SOFT-GRASS) TYPE

VEGETATION

Key species

Constant species: *Ranunculus repens* (creeping buttercup), *Oxalis acetosella* (wood-sorrel), *Agrostis tenuis* (common bent-grass), *Viola riviniana* (common violet), *Epilobium montanum* (broad-leaved willow-herb), *Holcus mollis* (creeping soft-grass), *Anthoxanthum odoratum* (sweet vernal-grass), *Betula* spp (birch), *Dryopteris filix-mas* (male fern), *Holcus lanatus* (Yorkshire fog)

Plot dominants: *Holcus mollis* (creeping soft-grass), *Pteridium aquilinum* (bracken), *Agrostis tenuis* (common bent-grass)

Selective species: *Succisa pratensis* (devil's-bit scabious), *Ranunculus repens* (creeping buttercup), *Veronica officinalis* (common speed-well), *Geranium sylvaticum* (wood cranesbill), *Campanula rotundifolia* (harebell), *Pseudoscleropodium purum*

Species groups: E, I, D, G, F, J, A (C, H, B)

Canopy and understorey species

Constant trees (birch, rowan)	*Constant saplings* (birch)
Constant shrubs (hazel)	*Trees (basal area)* birch

ENVIRONMENT

Geographical distribution
ES (WS, NW, Wa, NE)

Solid geology
Ign/Metam, Red s st (H, G)

Altitude (m)	Slope (°)	Rainfall (cm)	Soil pH	LOI
115 *(med)*	25 *(high)*	112 *(med)*	5·4 *(med)*	14·1 *(low)*

GENERAL DESCRIPTION

A very heterogeneous type with a high species complement, most closely related to types 29, 16 and 26. There is an average ground cover but with a high cover of rocks and bryophytes. The canopy is usually open with an understorey often present, but saplings also present in low density.

This type would probably be called herb-rich upland birch wood. The soils are mainly gleys and are usually flushed being often besides streams. The comparable phytosociological associations are probably Melico-Betuletum (KL pers comm) and Betula herb nodum McVean & Ratcliffe 1959.

Type 30. *Succisa pratensis/Holcus mollis* (devil's-bit scabious/creeping soft-grass) type
Ground flora species: *Blechnum spicant* (hard-fern), *Dactylorchis fuchsii* (common spotted orchid), *Dryopteris filix-mas* (male fern), *Agrostis tenuis* (common bent-grass)
Woody species: *Alnus glutinosa* (alder), *Corylus avellana* (hazel)

-   Absent
•   1 – 4
●   5 – 8
●   9 – 12
●   13 – 16

PLOT TYPE 31

TRIFOLIUM PRATENSE/HOLCUS LANATUS (RED CLOVER/YORKSHIRE FOG) TYPE

VEGETATION

Key species

Constant species: *Agrostis tenuis* (common bent-grass), *Holcus lanatus* (Yorkshire fog), *Trifolium repens* (white clover), *Cerastium holosteoides* (common mouse-ear chickweed), *Anthoxanthum odoratum* (sweet vernal-grass), *Plantago lanceolata* (ribwort).

Plot dominants: *Holcus lanatus* (Yorkshire fog), *Agrostis tenuis* (common bent-grass), *Anthoxanthum odoratum* (sweet vernal-grass), *Cynosurus cristatus* (crested dog's-tail), *Holcus mollis* (creeping soft-grass), *Lolium perenne* (rye-grass), *Deschampsia cespitosa* (tufted hair-grass)

Selective species: *Trifolium pratense* (red clover), *Trifolium repens* (white clover), *Cynosurus cristatus* (crested dog's-tail), *Lotus corniculatus* (birdsfoot trefoil), *Plantago lanceolata* (ribwort), *Lolium perenne* (rye-grass)

Species groups: G, I (H, D, J, A, E, F, C, B)

Canopy and understorey species

Constant trees	*Constant saplings*
—	—

Constant shrubs	*Trees (basal area)*
—	—

ENVIRONMENT

Geographical distribution
Wa (WS, NE, SW, NW, ME, ES)

Solid geology
Mill grit/Coal mea, Carb li/Mag li, Ign/ Metam (H, D, B)

Altitude (m)	*Slope (°)*	*Rainfall (cm)*	*Soil pH*	*LOI*
115 *(med)*	9 *(low)*	122 *(med)*	5·6 *(med)*	10·6 *(low)*

GENERAL DESCRIPTION

A type of infrequent occurrence within an average range of sites, with high heterogeneity and an average species complement, most closely related to types 15, 30 and 16. There is often a high ground cover, frequently grazed by cattle and sheep, but with a high cover of litter also. The canopy is usually very open, with few saplings or shrubs.

This type is mainly scrub or completely open glades. The soils are mainly gleys. The comparable phytosociological association is probably Lolio-Cynosuretum (Br-Bl et De L 1936) Tx 1937 or Centaureo-Cynosuretum Br-Bl et Tx 1952, but there are also some affinities with other grassland associations.

Type 31. *Trifolium pratense/ Holcus lanatus* (red clover/Yorkshire fog) type
Ground flora species: *Dactylis glomerata* (cock's-foot), *Holcus lanatus* (Yorkshire fog)
Woody species: *Quercus* spp (oak), *Ulex europaeus* (gorse)

-   Absent
- • 1 – 4
- • 5 – 8
- ● 9 – 12
- ● 13 – 16

PLOT TYPE 32

EPILOBIUM PALUSTRE/CAREX VESICARIA (MARSH WILLOW-HERB/BLADDER SEDGE) TYPE

VEGETATION

Key species

Constant species: *Angelica sylvestris* (wild angelica), *Filipendula ulmaria* (meadow-sweet), *Galium palustre* (marsh bedstraw), *Juncus effusus* (soft rush), *Salix* spp (willow), *Epilobium palustre* (marsh willow-herb), *Lychnis flos-cuculi* (ragged robin), *Mentha aquatica* (water mint), *Acrocladium cuspidatum*

Plot dominants: *Carex vesicaria* (bladder sedge), *Agrostis tenuis* (common bent-grass), *Holcus lanatus* (Yorkshire fog), *Juncus articulatus* (jointed rush), *Juncus effusus* (soft rush)

Selective species: *Naumburgia thyrsiflora* (tufted loosestrife), *Stachys palustris* (marsh woundwort), *Veronica scutellata* (marsh speedwell), *Epilobium palustre* (marsh willow-herb), *Scutellaria galericulata* (skull-cap), *Carex vesicaria* (bladder sedge)

Species groups: G, C (D, E, I, H)

Canopy and understorey species

Constant trees
(willow, alder, birch)

Constant saplings
—

Constant shrubs
—

Trees (basal area)
(willow)

ENVIRONMENT

Geographical distribution
WS, NW (NE)

Solid geology
Red s st, Silur/Ordov (D, G)

Altitude (m)	Slope (°)	Rainfall (cm)	Soil pH	LOI
51 *(low)*	3 *(low)*	140 *(high)*	5·5 *(med)*	24·1 *(low)*

GENERAL DESCRIPTION

A very infrequent type within a restricted range of site types, with medium heterogeneity and an average species complement, most closely related to types 14, 15 and 31. There is usually a high ground cover. The canopy is usually very open, with few saplings and no understorey.

This type would probably be called willow or alder scrub. The soils are gleys developed from recent alluvium, usually with areas of surface water. Because of the small number of plots in this type, it is not well-defined. The comparable phytosociological association is possibly Carici laevigatae-Alnetum Allorge 1922.

Type 32. _Epilobium palustre/ Carex vesicaria_ (marsh willow-herb/bladder sedge) type
Ground flora species: _Iris pseudacorus_ (yellow flag)
Woody species: _Salix_ spp (willow), _Alnus glutinosa_ (alder)

–	Absent
•	1 – 4
•	5 – 8
●	9 – 12
●	13 – 16